HANDBOOK OF MICROBIOLOGY

HANDBOOK OF
MICROBIOLOGY

By

P.S. Bisen
D.Sc.
Department of Microbiology
Barkatullah University
Bhopal 462 026

and

Kavita Verma
M.Sc.
Department of Microbiology
Barkatullah University
Bhopal 462 026

CBS

CBS Publishers & Distributors Pvt. Ltd.

New Delhi • Bengaluru • Chennai • Kochi • Kolkata • Mumbai
Hyderabad • Nagpur • Patna • Pune • Vijayawada

Handbook of
MICROBIOLOGY

ISBN: 978-81-239-0269-7

First Edition: 1994
Reprint: 1996, 1998, 2002, 2004, 2007, 2009, 2012, 2016, 2021

Published by **Satish Kumar Jain** and produced by **Varun Jain** for

CBS Publishers & Distributors Pvt Ltd

4819/XI Prahlad Street, 24 Ansari Road, Daryaganj, New Delhi 110 002, India.
Ph: 011-23289259, 23266861, 23266867
Fax: 011-23243014
Website: www.cbspd.com
e-mail: delhi@cbspd.com; cbspubs@airtelmail.in.

Corporate Office: 204 FIE, Industrial Area, Patparganj, Delhi 110 092
Ph: 011-4934 4934 Fax: 011-4934 4935
e-mail: publishing@cbspd.com; publicity@cbspd.com

Branches

- **Bengaluru:** Seema House 2975, 17th Cross, K.R. Road, Banasankari 2nd Stage, Bengaluru 560 070, Karnataka, India
  Ph: +91-80-26771678/79 Fax: +91-80-26771680 e-mail: bangalore@cbspd.com
- **Chennai:** 7, Subbaraya Street, Shenoy Nagar, Chennai 600 030, Tamil Nadu, India
  Ph: +91-44-26680620, 26681266 Fax: +91-44-42032115 e-mail: chennai@cbspd.com
- **Kochi:** 42/1325, 1326, Power House Road, Opp KSEB, Ernakulum, Kochi 682 018, Kerala, India
  Ph: +91-484-4059061-65,67 Fax: +91-484-4059065 e-mail: kochi@cbspd.com
- **Kolkata:** 6/B, Ground Floor, Rameswar Shaw Road, Kolkata-700014 (West Bengal), India
  Ph: +91-33-2289-1126, 2289-1127, 2289-1128 e-mail: kolkata@cbspd.com
- **Mumbai:** PWD Shed, Gala no 25/26, Ramchandra Bhatt Marg, Next to JJ Hospital Gate no. 2, Opp. Union Bank of India, Noorbaug Mumbai-400009, Maharashtra, India
  Ph: +91-22-66661880/89 e-mail: mumbai@cbspd.com

Representatives

• Hyderabad	0-9885175004	• Jharkhand	0-9811541605	• Nagpur	0-9421945513
• Patna	0-9334159340	• Pune	0-9623451994	• Uttarakhand	0-9716462459

Printed at Rashtriya Printers, Dilshad Garden, Delhi, India

Preface

"Handbook of Microbiology" as presented here is patterned after a series of experiments, discussions and deep thinking on the matter.

In recent years, microbiology has infiltrated into all branches of biology. Today, microbiological techniques are employed successfully for varied purposes as studying the pathogenicity of an organism, growing auxotrophs, for several industrial and biotechnological purposes, freeing plants of pathogens and other innumerable works for the welfare of mankind. The demonstration topics and the techniques described in the book reflect this diversity.

We believe that many biologists would like to use microbiological techniques in their research, but hesitate to do so because of lack of information. Inspite of several books on the physiology, biochemistry and molecular biology of microorganisms there exists no single book which can explain the metabolism of tiny creatures with a practical approach and which includes a brief text on the microbiological experiments useful for the workers in the allied fields at various levels.

A constant need for such a text material was widely felt which describes the methods in simple language and in bold outline supplemented with illustrations and various media composition and preparation of several stains.

Our basic aim in bringing out this book is to provide a training to the students of microbiology in handling and understanding microorganisms and also bringing them closer to the physiology of living cells.

Beginning with characteristic differences between eukaryotes and prokaryotes the book depicts the various methods employed to characterise and cultivate bacteria. Several staining procedures have been compiled at one place. The subject matter for various biochemical tests and their results are illustrated by colour plates and line diagrams.

A major section of the book deals with the occurrence, morphology and various biochemical assays for isolating and identifying a wide range of bacteria. The chapter has been made complete by giving the expected results and flowcharts to make identification easier till the species level.

Separate chapters on membrane filter technique and agglutination reactions include the various techniques in detail.

Classification of various fungi and experimental exercises with suitable illustrations and micrometery also forms a part of this book.

The cyanobacteria and viruses comprise the last two sections of the book. Every care is taken to include the minute details. Authors claim no originality for the methods described in the book, except for the fact that each method has been carried out in the laboratories before being included. It is, therefore, presumed that there would be no difficulty for any moderately equipped laboratory in the use of these methods.

The exercises have been formulated by keeping in mind the hazards of chemicals and only those exercises are given which minimise the exposure of students to potentially pathogenic organisms and hazardous chemicals.

We are grateful to Mr. M.P. Soni for beautiful illustrations. Additional thanks go to Mr. D. Trinad for his excellent typing of the manuscript.

This book is an attempt to help students to know their subject better and bring the various physiological relations closer to them. Every care has been taken to bring this volume in an exclusive, illustrative and easier form.

Constructive criticism and suggestions for further improvement of the book will be appreciated and gratefully acknowledged.

AUTHORS

Contents

1

Introduction

OUTLINE

• Distinguishing features of Prokaryotes and Eukaryotes • Classification of bacteria

In 1886, a German Zoologist E.H. Haeckal, placed the unicellular microorganisms that were typically neither plants nor animals in a third kingdom *Protista* (first two being Animalia and Plantae). The Protista includes bacteria, algae, fungi and protozoa. Haeckal's kingdom *Protista* left a question unanswered what criteria could be used to distinguish a bacterium from yeast?

By 1940 with the introduction of microscopy it was discovered that in some cells, e.g., typical bacteria, the nuclear substance was not enclosed by a nuclear membrane. This discovery, the absence of nuclear membrane-bound internal structure in one group of Protists (bacteria) and presence of bound structure in majority of the others (algae, fungi and protozoa), was a discovery of fundamental significance.

On the basis of above all characteristics, the former group has been named as Prokaryotes and the latter as Eukaryotes. These cell types have been characterised as in Table 1.

Table 1 Features Distinguishing Prokaryotic From Eukaryotic Organisms

FEATURE	PROKARYOTIC ORGANISMS	EUKARYOTIC ORGANISMS
1. Groups where found as unit of structure	Bacteria, Cyanobacteria, Mycoplasma	Algae, fungi, Protozoa, Plants and Animals
2. Size range of organism	1-2 by 1-4 μm or less	Greater than 5 μm in width or diameter
3. Genetic system location	Nucleoid, chromatin body, or nuclear material	Nucleus, mitochondria, chloroplasts
4. Structure of nucleus	(a) Not bound by nuclear membrane; One circular chromosome	(a) Bound by nuclear membrane; more than one chromosome

(Contd.)

FEATURE	PROKARYOTIC ORGANISMS	EUKARYOTIC ORGANISMS
	(b) Chromosomes does not contain histones; no mitotic division	(b) Chromosomes have histones; mitotic nuclear division
	(c) Nucleolus absent; functionally related genes may be clustered	(c) Nucleolus is present; functionally related genes are not clustered
5. Sexuality	Zygote is partially diploid known as merozygote	Zygote is diploid
6. Cytoplasmic nature and structure		
a. Cytoplasmic streaming	Absent	Present
b. Pinocytosis	Absent	Present
c. Gas vacuoles	May be present	Absent
d. Mesosome	Present	Absent
e. Ribosomes	70s distributed in cytoplasm	70s and 80s 80s are present on membranes as in endoplasmic reticulum, 70s present in mitochondria and chloroplast
f. Mitochondria	Absent	Present
g. Chloroplast	Absent	May be present
h. Golgi bodies	Absent	Present
i. Endo-plasmic reticulum	Absent	Present
k. Membrane bound true vacuoles	Absent	Present
7. Outer cell-structures		
a. Cytoplasmic membranes	Generally do not have sterols; it bears in it partial respiratory and photosynthetic unit	Sterols are present; independent photosynthetic and respiratory units
b. Cell wall	Peptidoglycan (murein or mucopeptide) as component	Peptidoglycan not present
c. Locomotor organelles	Simple fibril	Multifibrilled with "9 + 2" microtubules
d. Pseudopodia	Absent	Present in protozoans
8. Metabolic mechanisms	Varies : aerobic, anaerobic, fermentative, nitrogen fixing	Varies but glycolysis is the essential catabolic pathway
9. DNA base ratios as moles % of guanine + cytosine (G + C %)	28 to 73	About 40

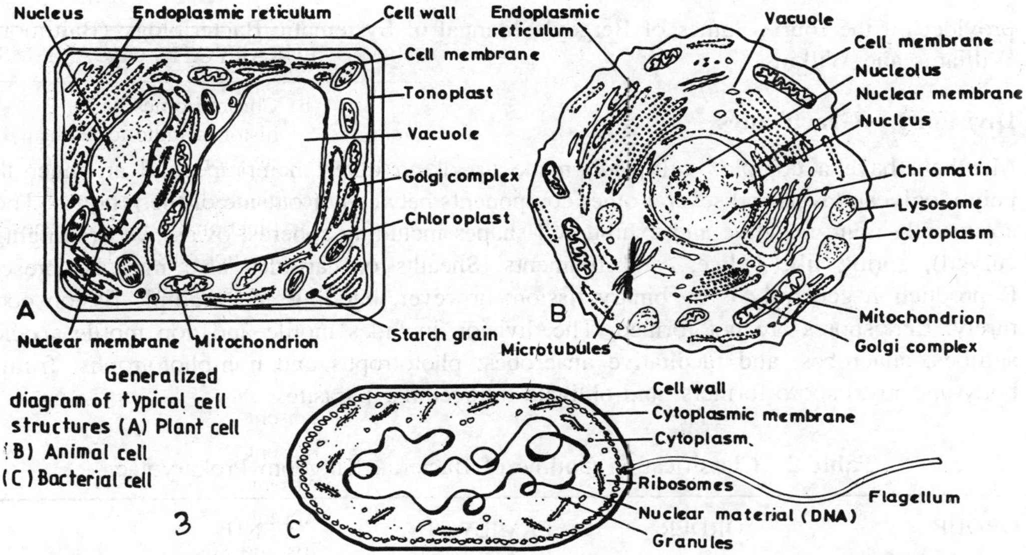

Generalized diagram of typical cell structures (A) Plant cell (B) Animal cell (C) Bacterial cell

3

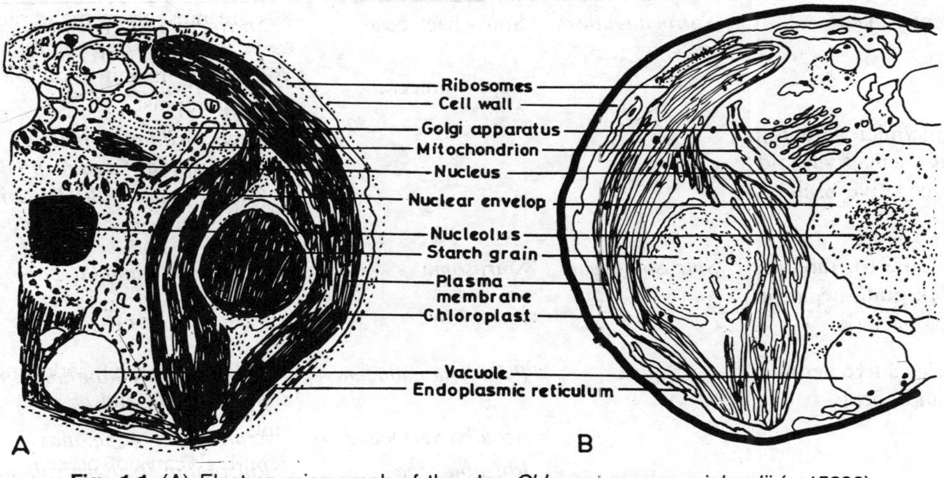

Fig. 1.1 (A) Electron micrograph of the alga *Chlamydomonas reinhardii* (× 15000), a eucaryotic cell (B) Schematic representation of (A).

Several classifications were given, of which Bergey's manual of systematic bacteriology has wide acceptance. It places all bacteria in kingdom *Prokaryotae* which is divided into 4 divisions.

Different systems of classification exist for fungi and algae. Taxonomy for viruses is still in progress.

BERGEY'S MANUAL

The classification scheme presented here is based on the information contained and/or to be provided in the four volumes of Bergey's Manual of Systematic Bacteriology (Baltimore : Williams and Wilkins).

Division I – Gracilicutes

Members have a complex cell wall, made up of an outer membrane and an inner thin peiptidoglycan layer, and several other components between or outside of these layers. These are usually gram-negative and exhibit cell shapes including spheres, ovals, rods (straight or curved), spring like helices, and filaments. Sheaths or capsules also may be present. Reproduction generally is by binary fission; however, budding and multiple fission occur rarely. Endospores are not formed. The division includes motile and non motile species; aerobes, anaerobes, and facultative anaerobes; phototrophs and non-phototrophs; fruiting body and myxospore formers; and obligate intracellular parasites.

Table 2 Classification Outline of Bacteria, Kingdom Prokaryotae

GROUP	ORDER	FAMILY	GENUS
The spirochaetes	*Spirochaetales*	Spirochaetaceae	*Spirochaeta Cristispira Treponema Borrelia*
		Leptospiraceae	*Leptospira*
Aerobic/microaerophilic motile, helical/vibrioid gram-negative bacteria			*Aquaspirillum Spirillum Azospirillum Oceanospirillum Campylobacter Bdellovibrio Vampirovibrio*
Nonmotile (or rarely motile), gram-negative curved bacteria	*Spirosomaceae*	*Spirosoma*	*Spirosoma Runella Flectobacillus*
Gram-negative aerobic rods and cocci		*Pseudomonadaceae*	*Pseudomonas Xanthomonas Frateuria Zoogloea*
		Azotobacteriaceae	*Azotobacter Azomonas*
		Rhizobiaceae	*Rhizobium Bradyrhizobium Agrobacterium Phyllobacterium*
		Methylococcaceae	*Methylococcus Methylomonas*
		Acetobacteriaceae	*Acetobacter Gluconobacter*
		Legionellaceae	*Legionella*
		Neisseriaceae	*Neisseria Moraxella Acinetobacter Kingella*
Facultatively anaerobic gram-negative rods		*Enterobacteriaceae*	*Escherichia Shigella Salmonella Citrobacter*

(Contd.)

GROUP	ORDER	FAMILY	GENUS
			Klebsiella Enterobacter Erwinia Serratia Hafnia Edwardsiella Proteus Providencia Morganella Yersinia Obesumbacterium Xenorhabdus Kluyvera Rahnella Cedecea Tatumella
		Vibrionaceae	*Vibrio Photobacterium Aeromonas Plesiomonas Pasteurella Haemophilus Actinobacillus*
Anaerobic gram-negative straight, curved, and helical rods		*Bacteriodaceae*	*Bacteroides Fusobacterium Leptotrichia Butyrivibrio Succinimonas Succinivibrio Anaerobiospirillum Wolinella Selenomonas Anaerovibrio Pectinatus Acetivibrio Lachnospira*
Dissimilatory sulfate or sulfur-reducing bacteria			*Desulfuromonas Desulfovibrio Desulfomonas Desulfococcus Desulfobacter Desulfolobus Desulfosarcina*
Anaerobic gram-negative cocci		Veillonellaceae	*Veillonella Acidaminococcus Megasphaera*
Chemolithotrophic		Nitrobacteraceae (Nitrifiers)	*Nitrobacter Nitrospira Nitrococcus Nitrosomonas Nitrosospira Nitrosococcus Nitrosolobus*
		Sulfur oxidizers	*Thiobacillus Thiomicrospira Thiobacterium Thiospira Macromonas*
		Obligate hydrogen oxidizers	*Hydrogenbacter*
		Siderocapsaceae (Metal oxidizers and depositers)	*Siderocapsa Naumanniella Ochrobium Siderococcus*
Anoxygenic photosynthetic bacteria	Purple bacteria	Chromatiaceae	*Erythrobacter Chromatium Thiocystis Thiospirillum*

(Contd.)

GROUP	ORDER	FAMILY	GENUS
			Thiocapsa Amoebobacter Lamprobacter Lamprocystis Thiodictyon Thiopedia
		Ectothiorhodospiraceae (Rhodospirillaceae) (Purple non-sulfur bacteria)	*Ectothirhodospira Rhodospirillum Rhodopseudomonas Rhodobacter Rhodomicrobium Rhodopila Rhodocyclus*
	Green bacteria	Chlorobiaceae (Green sulfur bacteria)	*Chlorobium Prosthecochloris Pelodictyon Chloroherpeton Symbiotic Consortia*
		Chloroflexaceae (Multicellular filamentous green bacteria)	*Chloroflexus Heliothrix Oscillochloris Chloronema*
		Others	*Heliobacterium*
Cyanobacteria			*Anabaena Chroococcus* etc.
Others	Prochlorales	Prochloraceae	*Prochloron*
Budding and/or appendaged bacteria	Prosthecate bacteria	(Budding bacteria)	*Hyphomicrobium Hyphomonas Pedomicrobium Filomicrobium Dicotomicrobium Tetramicrobium Stella Ancalomicrobium Prosthecomicrobium*
		(Nonbudding bacteria)	*Caulobacter Asticcacaulis Prosthecobacter Thiodendron*
	Nonprosthecate bacteria	(Budding bacteria)	*Planctomyces Pasteurella Blastobacter Angulomicrobium Gemmiger Ensifer Isosphaera*
		(Nonbudding bacteria)	*Isosphaera Gallionella Nevskia*
		Morphologically unusual budding bacteria involved in iron and manganese deposition	*Seliberia Metallogenium Caulococcus Kuznezovia*
		Others	*Spinate bacteria*

(Contd.)

GROUP	ORDER	FAMILY	GENUS
Sheathed bacteria			*Sphaerotilus Leptothrix Haliscominobacter Lieskeella Phramidiothrix Crenothrix Clonothrix*
Gliding, fruiting bacteria	Myxobacteriales	Myxococcaceae	*Myxococcus*
		Archangiaceae	*Archangium Cystobacter*
		Cystobacteriaceae	*Melittangium Stigmatella*
		Polyangiaceae	*Polyangium Nannocytis Chondromyces*
		Others	*Angiococcus*
Gliding, non-fruiting bacteria	Cytophagales	Cytophagaceae	*Cytophaga Sporocytophaga Capnocytophaga Flexibacter Microscilla Saprospira Herpetosiphon*
	Lysobacteriales	Lysobacteriaceae	*Lysobacteriaceae Lysobacter*
	Beggiatoales	Beggiatoaceae	*Beggiatoa Thioploca Thiospirillopsis Thiothrix Achromatium*
		Simonsiellaceae	*Simonsiella Alysiella*
		Leucothricaceae	*Leucothrix Pelonema*
		Pelonemataceae	*Archroonema Peloploca Desmanthus*
		Families and genera of uncertain affiliation	*Toxothrix Vitreoscilla Chitinophagen Desulfonema*
The Rickettsias and Chlamydias	Rickettsiales	Rickettsiaceae	*Rickettsia*[a] *Rohalimaea*[a] *Coxiella Ehrlichia*[b] *Cowdria Neorickettsia*[b] *Wolbachia Rickettsiella*[c]
		Bartonellaceae	*Bartonella Grahamella*
		Anaplasmataceae	*Anaplasma Aegyptianella Haemobartonella Eperythrozoon*
	Chlamydiales	Chlamydiaceae	*Chlamydia*

a – Tribe Rickettsieae b – Tribe Ehrlichieae c – Tribe Wolbachleae

Division II – Firmicutes

Members here have a cell wall with a thick peptidoglycan layer that is characteristic of gram-positive organisms. These prokaryotes generally stain gram-positive and exhibit shapes including spheres, rods, and filaments. Branching of the rods and filaments does occur.

Reproduction is generally by binary fission. Endospores or spores on hyphae are formed by some species. Most organisms are chemosynthetic heterotrophs and not photosynthetic. The division includes, aerobes, anaerobes, facultative anaerobes, asporogenous and sporogenous species and the actinomycetes together with related forms.

GROUP	FAMILY	GENUS
Gram-positive cocci	Micrococcaceae	*Micrococcus Stomatococcus Planococcus Staphylococcus*
	Deinococcaceae	*Deinococcus* "Pyogenic" *Streptococci* "Oral" *streptococci* "Lactic" *streptococci* and *enterococci Leuconostoc Pediococcus Aerococcus Gemella Peptococcus Peptostreptococcus Reminococcus Coprococcus Sarcina*
Endospore-forming gram-positive rods and cocci		*Bacillus Sporolactobacillus Clostridium Desulfotomaculum Sporosaricina Oscillospira*
Regular, nonsporing gram-positive rods		*Lactobacillus Listeria Erysipelothrix Brochothrix Renibacterium Kurthia Caryophanon*
Irregular, nonsporing gram-positive rods		Animal and saprophytic *Corynebacteria* Plant pathogenic *Corynebacteria Gardnerella Arcanobacterium Arthrobacter Brevibacterium Curtobacterium Caseobacter Microbacterium Aureobacterium Cellulomonas Agromyces Arachnia Rothia Propionibacterium Eubacterium Acetobacterium Lachnospira Butyrivibrio Thermoanaerobacter Actinomyces Bifidobacterium*
Mycobacteria	Mycobacteriaceae	*Mycobacterium*
Nocardioforms		*Nocardis Rhodococcus Nocardioides Pseudomonocardia Oerskovia Saccharopolyspora Micropolyspora Promicromonospora Intrasporangium*
Actinomycetes dividing in more than one plane		*Geodermatophilus Dermatophilus Frankia Tonsilophilux*
Sporangiate actinomycetes		*Actinoplanes* (including Amorphosporangium) *Streptosporangium Ampullariella Spirillospora Pilimelia Dactylosporangium Planomonopora*
Streptomycetes and their allies		*Streptomyces Streptoverticillium*

(Contd.)

GROUP	FAMILY	GENUS
		Actinopycnidium[d] *Actinosporangium Chainia*[d] *Elytrosporangium Microcellobosporia*
Other conidiate genera		*Actinopolyspora Actinosynnema Kineospora Kitasatospora Microbispora Micromonospora Microtetrospora Saccharomonospora Sporichthya Streptoalloteichus Thermomonospora Actinomadura Nocardiopsis Excellospora Thermoactinomyces*

d – The last five genera may be merged with the streptomycetes

Division III – Tenericutes

Members of this division lack cell wall and do not synthesize peptidoglycan precursors. These prokaryotes are enclosed by a unit membrane – the plasma membrane – and stain gram-negative. The cells are highly pleomorphic and may appear as large irregular vesicles to filamentous forms with branching projections. Many resemble naked L-forms that can be generated from various bacterial species in other divisions. Reproduction may occur by budding, fragmentation, and/or binary fission. Most species are nonmotile, but some show a type of gliding movement. The organisms in the division generally require complex media for growth and the inclusion of cholesterol and long-chain fatty acids. Growth on solid media is characterized by the formation of "Fried egg" colonies. The division includes saprophytes and pathogens of animals, plants, and tissue cultures. All members are completely resistant to betalactam antibiotics.

GROUP	ORDER	FAMILY	GENUS
The mycoplasmas	Mycoplasmatales	Mycoplasmataceae	*Mycoplasma Ureaplasma*
		Acholeplasmataceae	*Acholeplasma*
		Spiroplasmataceae	*Spiroplasma*
		Other Genera	*Anaeroplasma Thermoplasma Mycoplasmalike* Organisms of plants and invertebrates

Division IV – Mendosicutes

Most members have some type of cell wall that does not contain typical peptidoglycan components. The composition of cell walls may include large protein molecules or a variety of polysaccharides. The species in the division are either gram-positive or gram-negative and exhibit cell shapes including cocci, rods, filaments, and those resembling the mycoplasma (cells lacking cell walls). Many organisms are motile, endospores or other resting forms have not been reported. Most species are strict anaerobes. Members of the division include organisms that are ecologically and metabolically diverse and capable of living in extreme environments.

GROUP	ORDER	FAMILY	GENUS
Archaeobacteria	Methanobacteriales	Methanobacteriaceae Methanothermaceae	*Methanobacterium*, *Methanobrevibacter* *Methanothermus*
	Methanococcales	Methanococcaceae	*Methanococcus*
	Methanomicrobiales	Methanomicrobiaceae	*Methanogenium* *Methanomicrobium* *Methanospirillum*
		Methanosarcinaceae	*Methanosarcina* *Methanococcoides* *Methanolobus*
		Methanoplanaceae	*Methanoplanus*
		Other genera	*Methanothrix*
	Halobacteriales	Halobacteriaceae	*Halobacterium Halococcus*
	Thermoplasmales	Theroplasmaceae	*Thermoplasma*
	Sulfolobales	Sulfolobaceae	*Sulfolobus*
	Thermoproteales	Thermoproteaceae	*Thermoproteus Thermofilum*
		Desulfurococcaceae	*Thermococcus* *Desulfurococcus* *Thermodiscus Pyrodictium*
Endosymbionts			
Endosymbionts protozoa			*Holospora Caediabacter* *Pseudocaedibacter Lyticum* *Tectibacter*
Endosymbionts of insects			*Blattabacterium*
Endosymbionts of fungi and invertebrates other than arthropods			

2

Bacteria : General Account

OUTLINE

• Size • Shape • Flagellation • Cell structure • Spores and cysts

Bacteria are ubiquitous in their occurrence. They occur in air, on the land, in lakes, in oceans, in the intestinal tracts of man and animals. They eat dead organic matter. Few even cause disease in men, animals and plants.

Morphology and Structure

(1) **Size** Bacteria are small living organisms. The diameter of a bacterial cell varies from 0.1 μ to 20 μ and in rod forms the length is between 0.2 μ to 80 μ.

(2) **Shape** On the basis of their general shape, bacteria are classified as follows :

(i) Bacillus Rod shaped (pl. bacilli)

(ii) Coccus Spherical (pl. cocci)

(iii) Spirillum Spirally twisted (pl. spirilli)

Coccus bacteria are of following types :

A Micrococcus When there is a single cell

B Diplococcus When there are two cells

C Tetracoccus When there are four cells

D Streptococcus Cells in chains

E Staphylococcus Cells in clusters like grapes

F Sarcina Many cells forming a cuboidal figure

(3) **Flagellation** Some species posses flagella while others are non flagellated and move by the effect of Brownian movements.

A bacterial flagellum consists of a single proteinaceous fibril arising from a blepharoplast granule in the peripheral part of the cytoplasm. In some species the flagellum consists of a core and a sheath, several flagella may remain in a loose bundle and function as a single flagellum or a flagellum may consist of several strands each comparable to a single flagellar cone with a single basal granule.

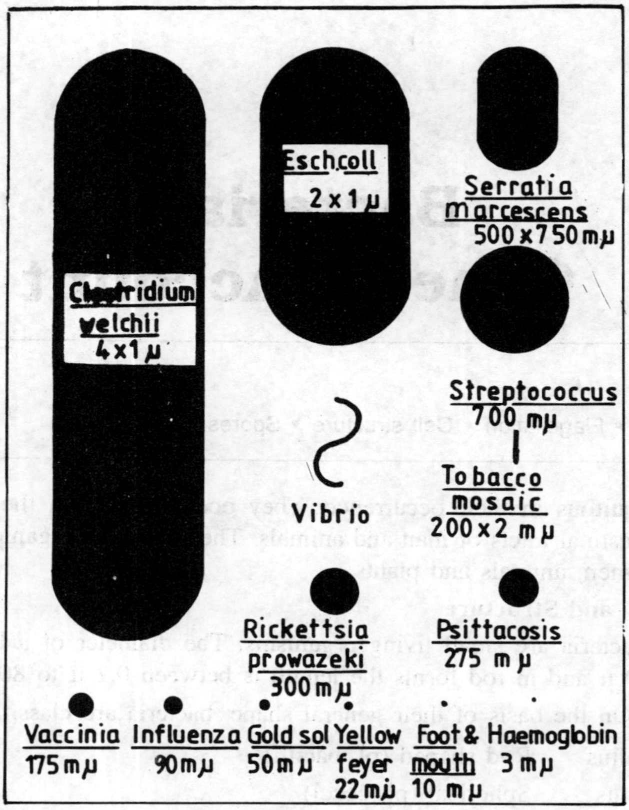

Fig. 2.1.

Depending upon the types of flagella the bacteria are classified into several categories as follows :

1. *Atrichous* Without flagella
2. *Monotrichous* With a single polar flagellum
3. *Lophotrichous* With a tuft of flagella at one pole only
4. *Amphitrichous* With a tuft of flagella at each of the two poles
5. *Peritrichous* With several flagella originating at several points over the entire margin

Cell Structure

A bacterial cell is prokaryotic in its construction, the chromatin material is neither concentrated in nucleus nor bounded by a nuclear membrane. The well defined chromosomes are not present. Each bacterium has a prominent cell wall. Some are surrounded by a viscous

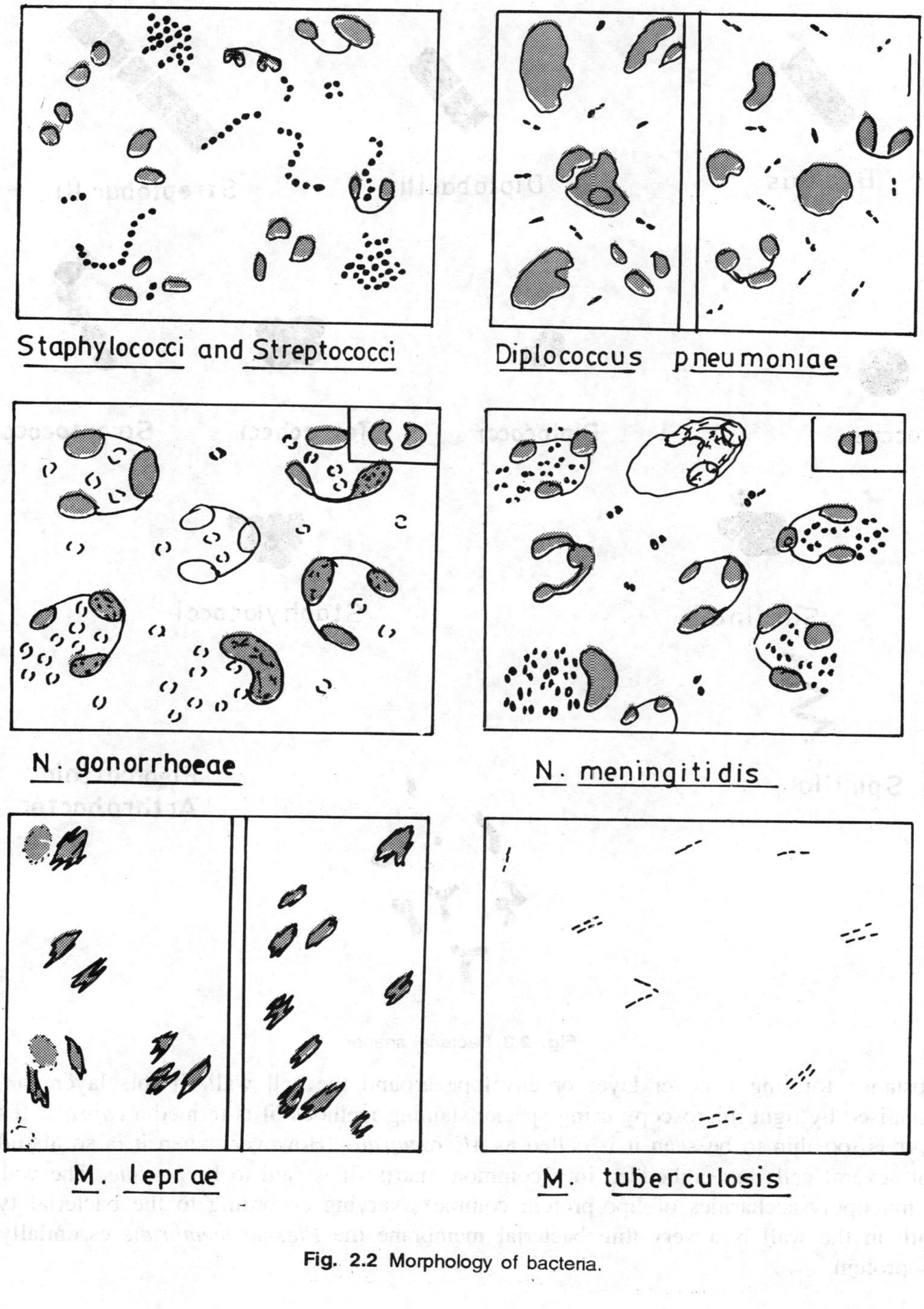

Fig. 2.2 Morphology of bacteria.

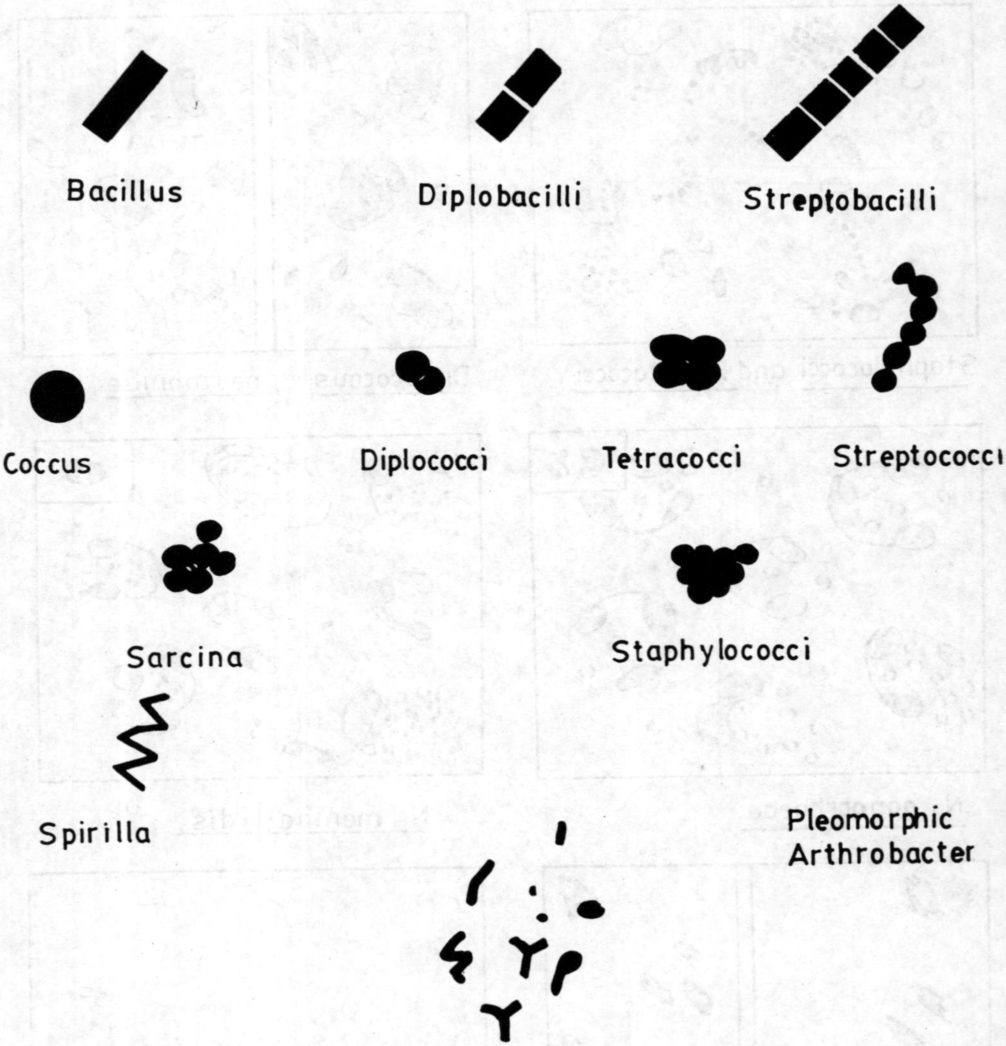

Fig. 2.3 Bacterial shapes.

substance forming a cover layer or envelope around the cell wall. If this layer can be visualised by light microscopy using special staining methods, it is termed a *capsule*. If this layer is too thin to be seen it is called as *Microcapsule*. However, when it is so abundant that several cells are embedded in a common matrix, it is said to be a *slime*. The wall is of mucopolysaccharides of lipo-protein complex, varying according to the bacterial type. With in the wall is a very thin bacterial membrane the *Plasma membrane* essentially of lipoprotein.

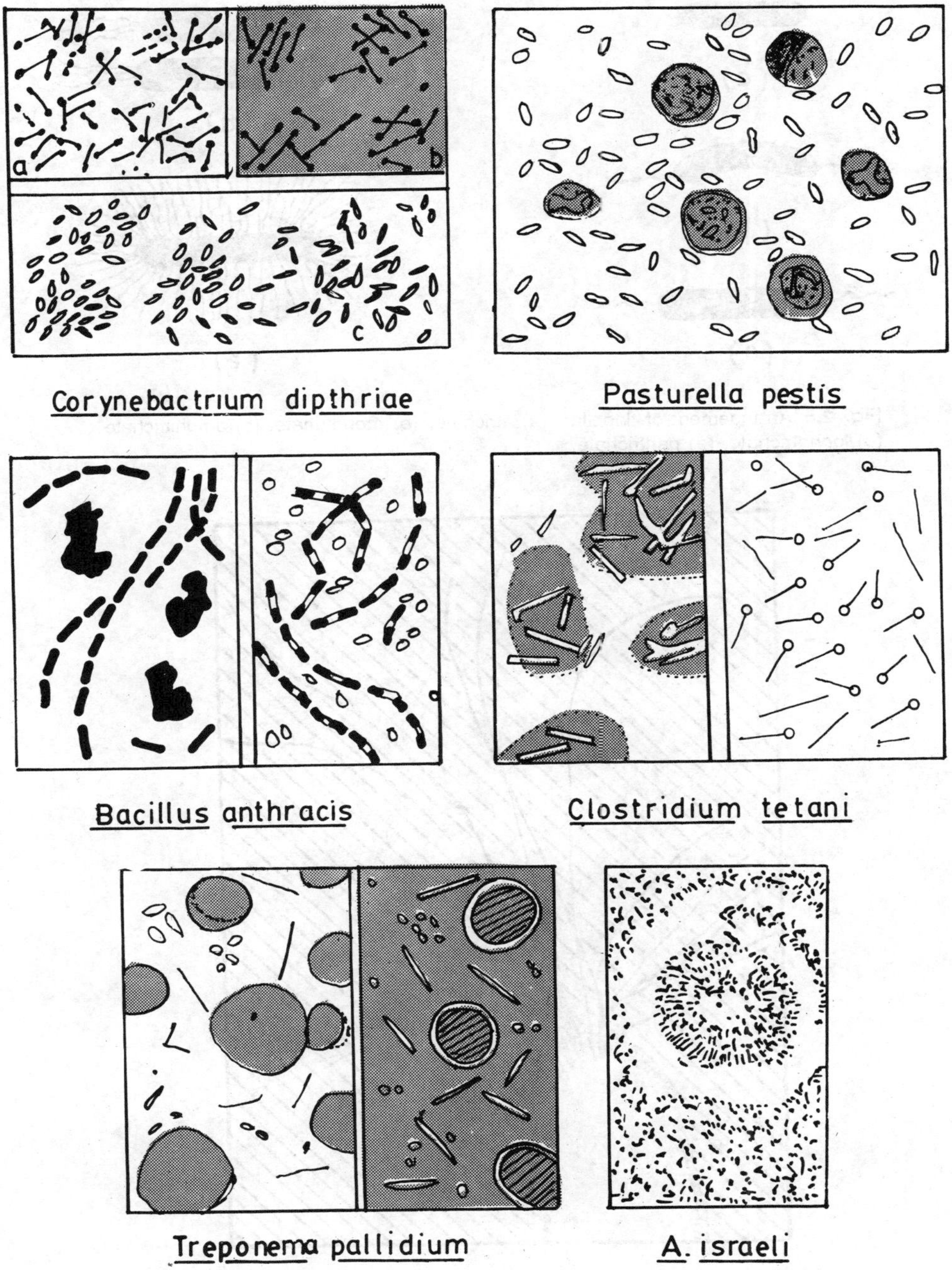

Fig. 2.4 Morphology of bacteria.

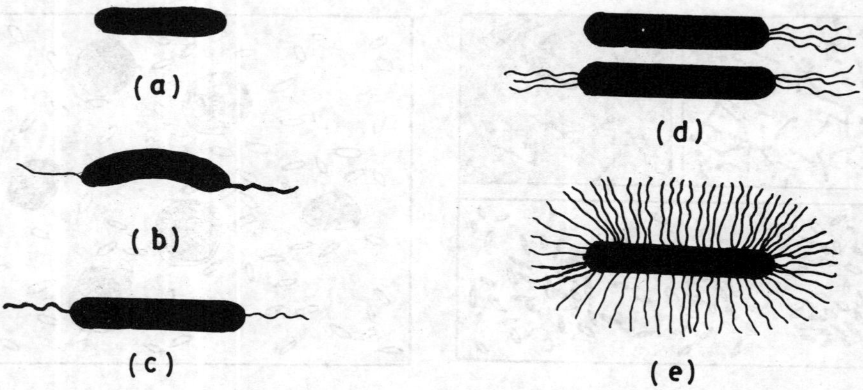

Fig. 2.5 Arrangement of flagella : (a) atrichate, (b) monotrihate, (c) amphitrichate, (d) lophotrichate, (e) peritrichate.

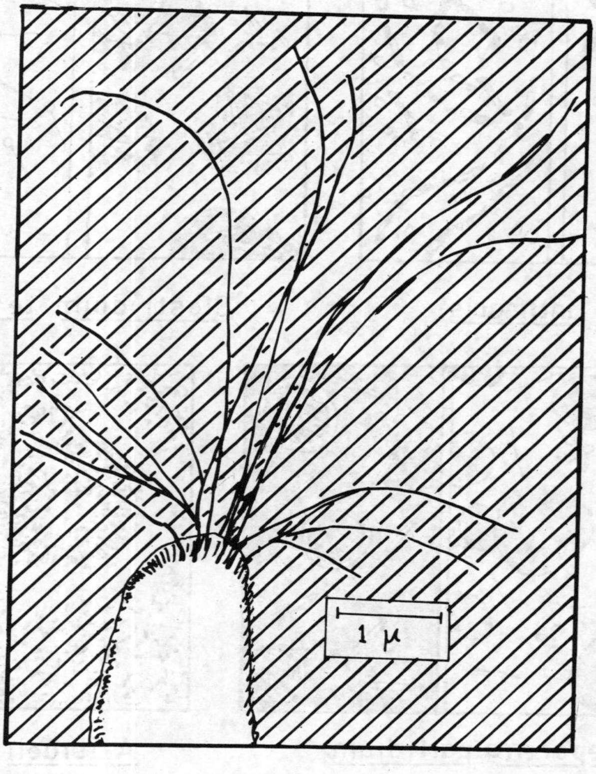

Fig. 2.6 Electron micrograph of *Spirillum serpens* showing lophotrichous flagellation, × 9000. The flagella are arising from the cytoplasmic membrane through the cell-wall.

Cytoplasm is granular mitochondria, plastids and other membrane bound organelles are lacking. However, the plasma membrane is invaginated at several places. A variety of enzymes, cytochromes etc. are located on the plasma membrane. Ribosomes occur freely in the cytoplasm. Volutin granules are also found.

In photosynthetic bacteria pigments are contained in membrane bound vesicles, situated near plasma membrane and often found continuous with it.

Volutin granules are composed of polyphosphate. Poly β hydroxybutyrate, a polymer found in aerobic bacteria under high carbon, low nitrogen condition and can be stained by lipid soluble dye Nile blue. Volutin granules acquire stain with methylene blue. Glyogen granules are known to acquire stain by iodine and look brown.

Bacterial cells lack nucleus but at the centre of the cell DNA is confined. DNA is single circular molecule in which all the genes are linked.

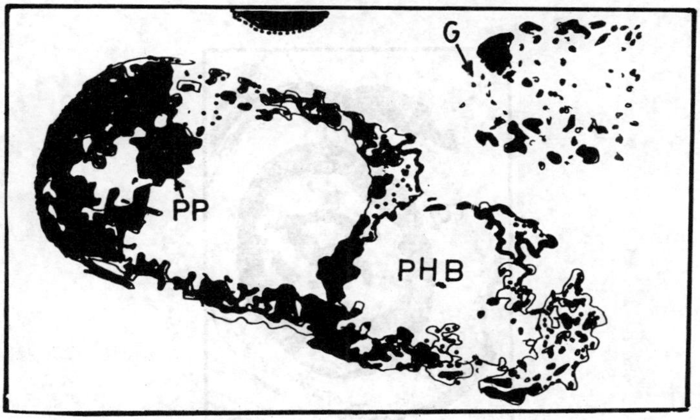

Fig. 2.7 Thin section of *Pseudomonas pseudoflava* showing polyphosphate (volutin) granules (PP), poly-B-hydroxybutyrate granules (PHB), and glycogen-like granules (G).

Spores and Cysts

Certain species of bacteria produce spores; either within the cells or external to it : the former being *endospores* and the latter referred to as *exospore*. The spore is a metabolically dormant form which, under appropriate conditions can undergo and outgrow to form a vegetative cell.

Endospores are thick walled, highly refractile and are produced (one per cell) by *Bacillus, Clostridium, Sporosarcina, Thermoactionomycetes* and a few other genera.

The shapes vary from elliptical and central (*Bacillus*) to spherical with terminal location (*Clostridium*) even ovoid but sub-terminally placed (*C. subterminale*).

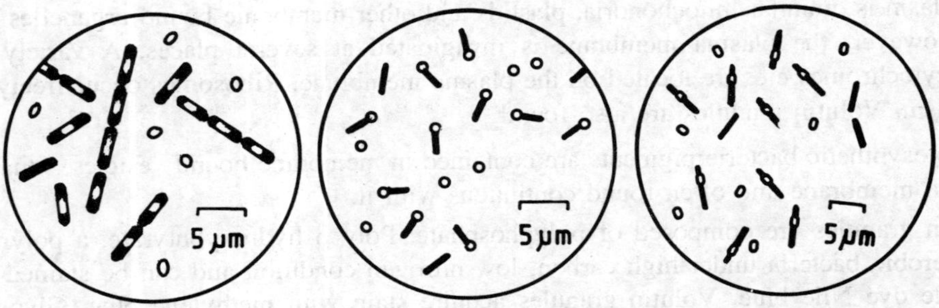

Fig. 2.8 Location of spores.

Fig. 2.9 Fine structure of an *Azotobacter* cyst. The exosporium (Ex) and the two layers of exine (CC_1 and CC_2 are visible. A nuclear region (Nr) and a cytoplasmic region containing ribosomes are observable within the central body.

Endospores are heat resistant and contain large amount of DPA (dipicolinic acid) a calcium complex.

Expospores are seen in cells of *Methylosinus* produced by budding.

Cysts are dormant, thick walled, desiccation-resistant forms and develop during differentiation of vegetative cell which can germinate under suitable conditions, e.g., *Azotobacter*.

3

Characterisation and Cultivation of Bacteria

OUTLINE

• Characterisation and cultivation of bacteria • Media preparation • Role of oxygen, temperature and pH • Cultivation of bacteria • Pure culture techniques

Nutritional requirements of bacteria vary greatly due to which there are vast differences in the chemical composition of the media used in the laboratory. Bacteria also exhibit the differences with respect to the physical conditions favouring the growth.

Bacteria can be classified into nutritional types as follows :

Table 3 Nutritional Types of Bacteria in Terms of Energy Sources and Hydrogen Donors

Energy Source and Hydrogen Donor	Nutritional Type
A. **Energy source**	
(i) Radiation (light)	Phototrophs
(ii) Oxidation of organic or inorganic compounds	Chemotrophs
B. **Hydrogen donor**	
(i) Reduced inorganic H donors (NO_2, NO_3, H_2S, S_2O_3, Fe^{2+}, CO, H_2)	Lithotrophs
(ii) Organic compounds	Organotrophs

Besides this depending on the need of oxygen for bacteria to survive, they have been divided into

(i) Aerobic

(ii) Facultatively Aerobic

(iii) Anaerobic

Aerobic bacteria grow on the surface of media, anaerobic bacteria accommodate themselves at the bottom. The facultatives form a suspension.

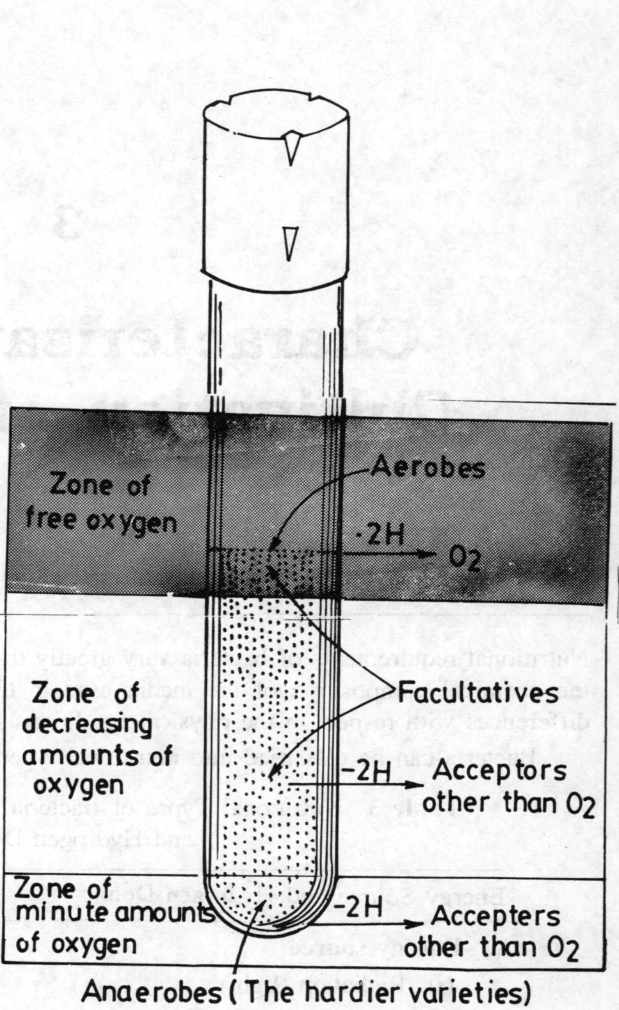

Fig. 3.1 (a) Relation of oxygen concentration in a solid medium to types of microbial growth.

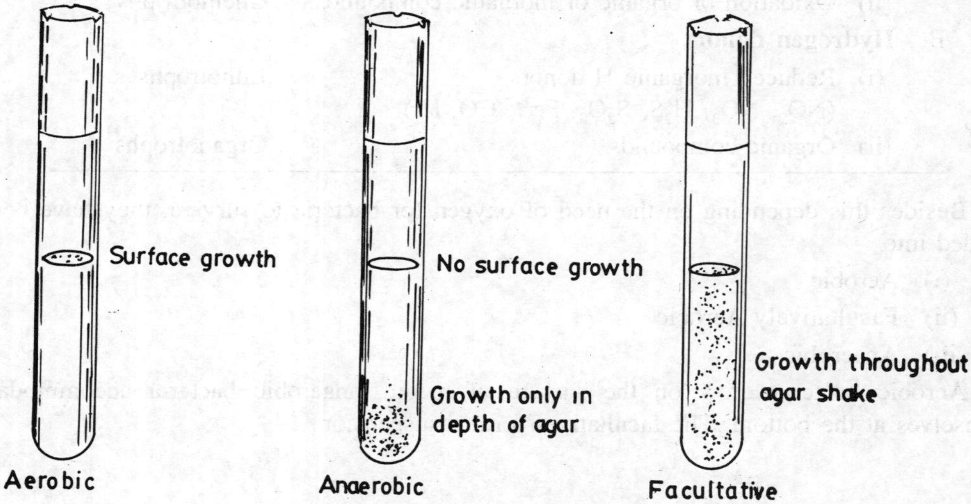

Addition of a dried polysaccharide agar, derived from a red algae is made when solid media is required. Agar has the property to go into solution (1.5-2%) at 100°C and solidify at 45°C.

Many special purpose media have been designed which facilitate recognition, enumeration and isolation of certain bacterial types.

(i) *Selective media :* It provides nutrients that enhance the growth and predominance of a particular type of bacterium and may inhibit other cell types.

(ii) *Differential Media :* Certain reagents or supplements when incorporated into culture media, may allow differentiation of several bacterial kinds.

Preparation of Media

1. Each ingredient or the completely dehydrated medium is dissolved in the appropriate volume of distilled water.
2. The pH of the fluid medium is determined by a pH meter and adjusted if necessary.
3. If a solid medium is required agar is added and the medium is boiled to dissolve the agar.
4. The medium is dispensed into tubes or flasks.
5. Media is sterilised by autoclaving (15 lbs pressure maintained for 20 mins). Some ingredients which are heat labile are sterilised by filteration.

ROLE OF OXYGEN, TEMPERATURE AND pH

Oxygen

The strong oxidising nature of oxygen, makes it a suitable terminal electron acceptor for the energy yielding process – the *respiration.* However, it may act as a toxic substance as in anaerobic organisms. Aerobic and facultative ones have developed protective mechanisms that greatly mitigate this toxicity. In anaerobes following factors are among those that have been implicated in oxygen toxicity.

(a) *Inactivation of enzymes by oxygen :* Molecular oxygen oxidises reduced groups (e.g., thiol –SH, or enzymes) thus rendering them inactive.

(b) *Derivatives of Oxygen :* Derivatives like Super-oxide (O_2^-) radicals can inactivate vital cell components.

$$O_2 + O^- \longrightarrow O_2^- \qquad ...(1)$$

Hydrogen peroxide generated by (O_2^-) is another poison.

$$2O_2^- + 2H^+ \longrightarrow O_2 + H_2O_2 \qquad ...(2)$$

The energized derivative of oxygen known as singlet oxygen (Δ_g) O_2, is a powerful oxidising agent highly toxic to cells.

In aerobes and facultative organisms, an enzyme *Superoxide dismutase* eliminates super oxide radicals as in reaction (2). The hydrogen peroxide produced by this reaction can in turn be dissipated by *Catalase* and *Peroxidase* enzymes.

$$2H_2O_2 \xrightarrow{\text{Catalase}} 2H_2O + O_2$$

$$H_2O_2 + \text{reduced substrate} \xrightarrow{\text{Peroxidase}} 2H_2O + \text{oxidised substrate}$$

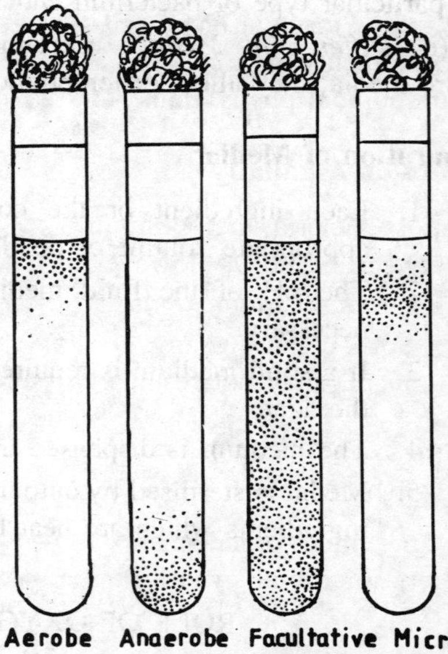

Fig. 3.2 Schematic illustration of the growth of bacteria in deep agar tubes, showing differences in response to atmospheric oxygen.

Temperature

The temperature that allows most rapid growth during a short period of time (12 to 24 h) is known as the optimum growth temperature.

Table 4 below shows optimum growth temperature for many bacteria

	Temperature of Growth °C		
	Minimum	Optimum	Maximum
Vibrio psychoerythrus	0	15	19
Pseudomonas fluorescence	4	25-30	40
Staphylococcus aureus	6.5	30-37	46
Corynebacterium diptheriae	15	37	40
Streptococcus thermophilus	20	40-45	50
Thermus aquaticus	40	70-72	79

On the basis of their temperature relationships, bacteria are divided into 3 main groups.

(1) *Psychrophiles :* Bacteria able to grow at 0°C or lower.

(2) *Mesophiles :* Grow best with in a temperature range of approximately 25 to 40°C.

(3) *Thermophiles :* Grow best at temperature above 45°C.

However, it is important to note that a bacterial species may not manifest the same characteristics in every detail when grown at different temperatures. For example : *Serratia marcescens* forms a blood red to orange pigment when cultured at 25°C but produces little or no pigment when cultured at 37°C. Similarly, *Lactobacillus planatarum* does not require the amino acid phenylamine for growth when cultured at 25°C but does require it at 37°C.

Acidity or Alkalinity

For most bacteria, the optimum pH for growth lies between 6.5 and 7.5 and limits generally lie between 5 and 9. However, a few bacteria prefer more extreme pH values for growth, e.g., *Thiobacillus thiooxidans* has an optimum pH of 2 to 3.5.

Before preparing any growth media pH should be adjusted, for any deviation from optimum pH may not result in accurate result.

CULTIVATION OF BACTERIA

In nature, micro organisms exist as mixed populations of many widely differing types. Many different microbial species normally inhabit various parts of our bodies, such as oral cavity,

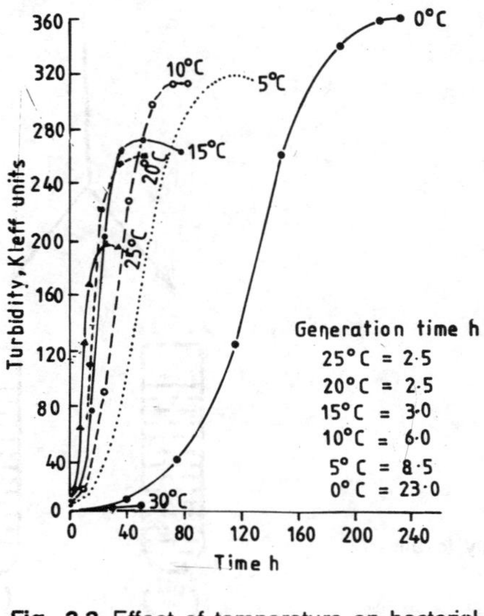

Fig. 3.3 Effect of temperature on bacterial growth.

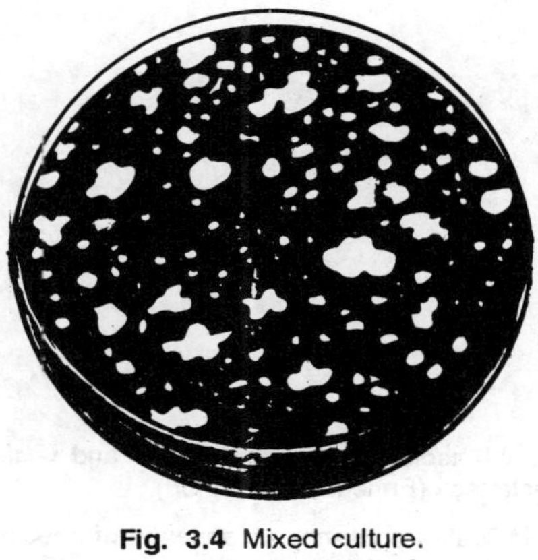

Fig. 3.4 Mixed culture.

intestinal tract and the skin. A single sneeze may therefore disperse from 10,000 to 100,000 bacteria. One gram of faeces contain 10^{11} bacteria. A study of microorganisms in these habitats require knowledge of the specific microbes present. This, in turn, requires unravelling of mixed population into pure culture of separate, different species.

3 kinds of selective methods are employed to obtain pure cultures.

(i) Chemical
(ii) Physical
(iii) Biological

(i) Chemical Methods of Selection

(a) *Use of specific carbon or nitrogen source :* The method is an example of enrichment media preparation. Here, for example if we wish to select nitrogen fixing organism, N_2 gas can be supplied as sole nitrogen source. One classical example deals with the isolation of *Pseudomonas* where supply of sodium benzoate as sole carbon source ensures the development of only desired bacterial colonies.

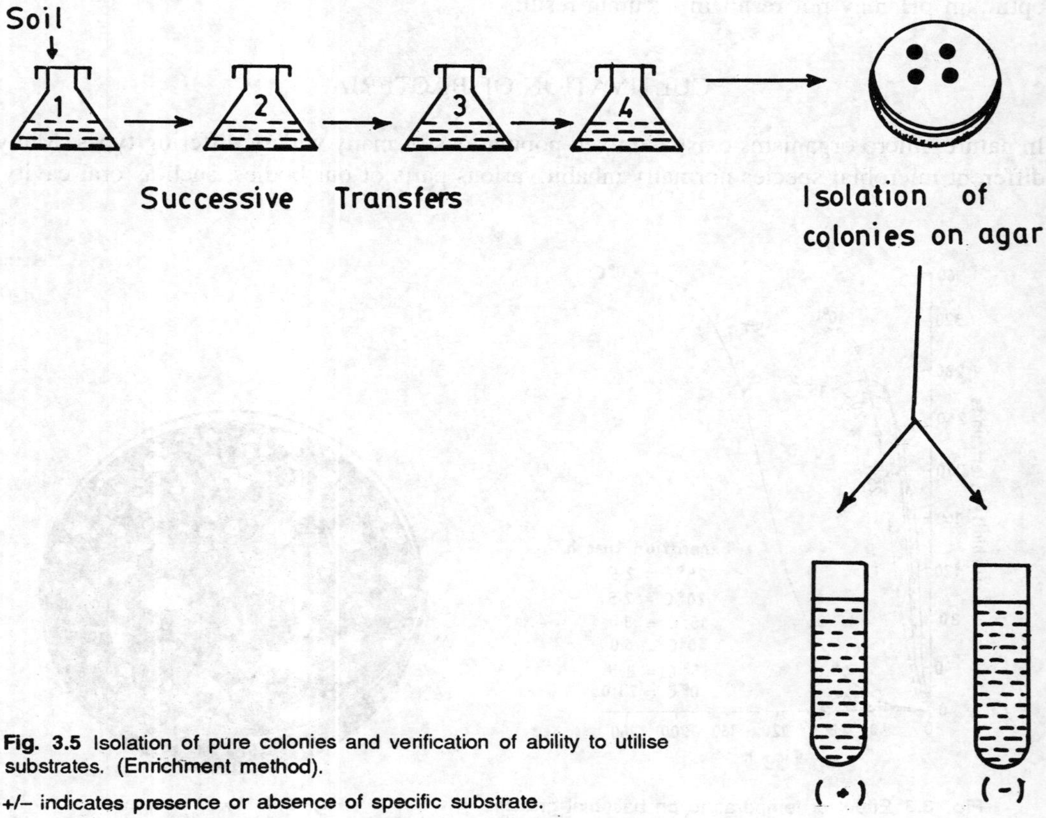

Fig. 3.5 Isolation of pure colonies and verification of ability to utilise substrates. (Enrichment method).

+/– indicates presence or absence of specific substrate.

(b) *Use of dilute media :* Certain bacteria like *Caulobacter* are capable of growing with very low levels of carbon and nitrogen source. In order to isolate such types, a media containing 0.01% peptone is employed since other bacteria are enable to grow in such dilute conditions.

(c) *Addition of toxic chemicals :* Gram-negative bacteria can grow in presence of low concentrations of various dyes that inhibit the growth of gram positive species. A medium containing crystal violet dye plus sodium de-oxy cholate units allow gram-negative intestinal bacterial to grow but will inhibit most other kinds. Use of Mac Conkey agar which is used to select for gram-negative intestinal pathogens such as *Salmonella* and *Shigella* is based on above principle.

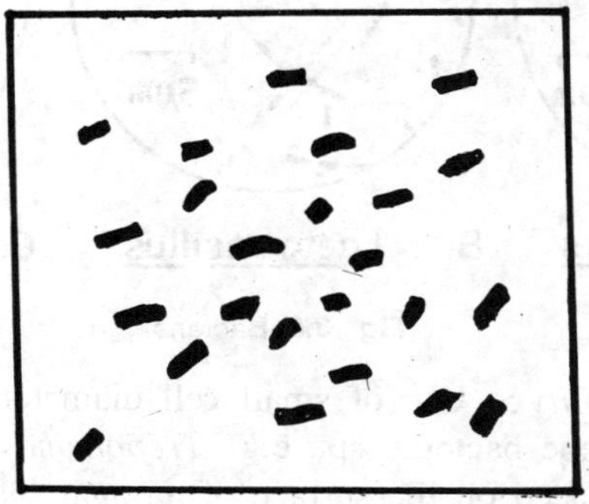

Fig. 3.6 Culture of *Salmonella typhosa.*

(ii) Physical Methods of Selection

(a) *Heat treatment :* To select for endospore forming bacteria, e.g., *Bacillus, Clostridium* a mixed culture is heated to 80°C, 10 min before being used to inoculate culture media. Vegetative cells will be killed by this treatment but endospores will survive.

Fig. 3.7 Electron micrograph of an endospore of *Bacillus cereus.*

(b) *pH and incubation temperature :* As discussed in previous chapter, to select for psychrophillic bacteria cultures are incubated at low temperature, e.g., 0 to 5°C. On the other hand, a hunt for thermophiles, is achieved at an incubation temperature of 55°C. Similarly, to select for acid tolerant bacteria a low pH medium can be used e.g. *Lactobacilli* (pH 5.35) while alkali tolerant like *Vibrio cholerae* obtained from stool sample thrives at pH 8.5.

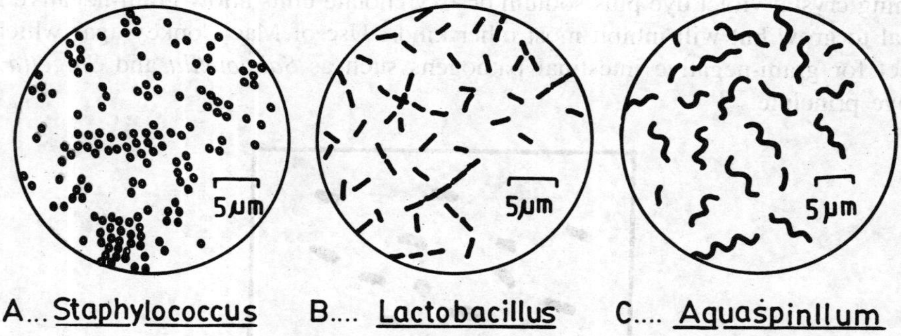

Fig. 3.8 Bacteria.

(c) *Cell size and motility :* Use of small cell diameter and ability to migrate helps smooth identification in some bacterial sp., e.g., *Treponema* sp., membrane filter having a pore size of 0.15 μm is placed on the surface of an agar plate and gingival scrapings are placed on the filter. The small *Treponemes* penetrate the pores of filter and reach the underlying surface. The motility feature help them to migrate away from filter.

(iii) Biological Methods of Selection

A disease-producing sp. occurring in a mixed culture can often be selected by taking advantage of its pathogenic properties. For example, a sputum having *S. pneumoniae* is ordinarily contaminated by many bacterial species. However, laboratory mice are extremely susceptible to infection by *S. pneumoniae*, and if the sputum sample is injected into a mouse the pathogen will multiply extensively.

Before commencing to the techniques related with pure culture, a knowledge of certain microbiological terms is essential.

Essentially, all culture media are of 2 forms : liquid or broth media and solid or agar media.

Agar enters into solution at 97-100°C and solidifies at about 45°C. It is not included as food, but only as an hardening agent.

The technique which ensures prevention of the entry of undesired microbes is aseptic technique which is followed at every step of cultivation. Since any medium immediately after preparation contains microorganisms from its ingredients and from the surfaces of the utensils for glasswares, it must be sterilized by heating or other methods. Prior to sterilization a tube containing the medium is plugged with cotton or plastic closures. This prevents the entry of new contaminants but permits free interchange of gases or air.

Inoculation is the method by which the desired microbe is transferred to the sterilized media. This is achieved by heating the inoculation loop to redness by flaming immediately before and after making the transfer. The flaming destroys any living forms on the surface of needle or loop (Fig. 3.9).

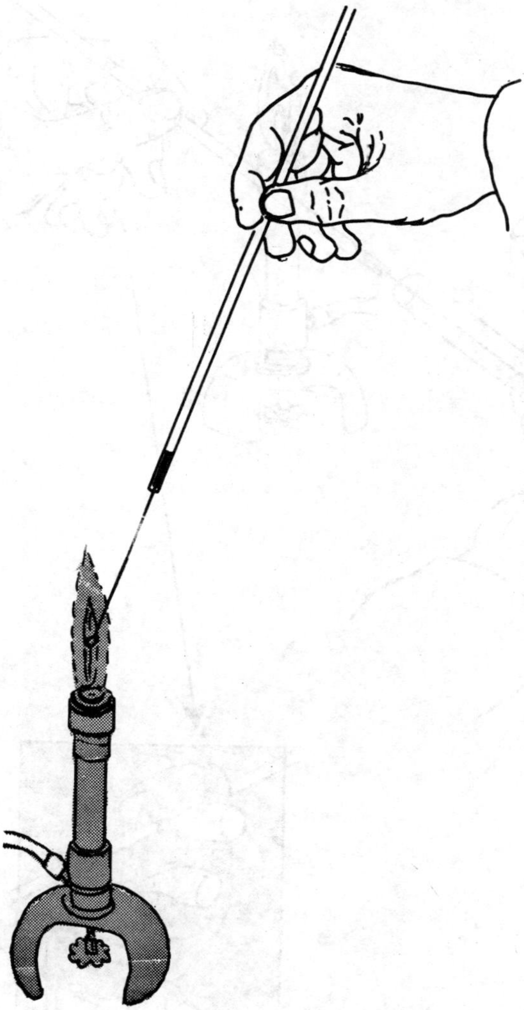

Fig. 3.9 Flame the inoculating loop before and after inoculation.

Following inoculation the bacterial culture is stored or incubated in an environment suitable for growth. The mass of daughter cells visible to naked eye either as cloudiness or turbidity in liquid broth or as an isolated colony on solid media.

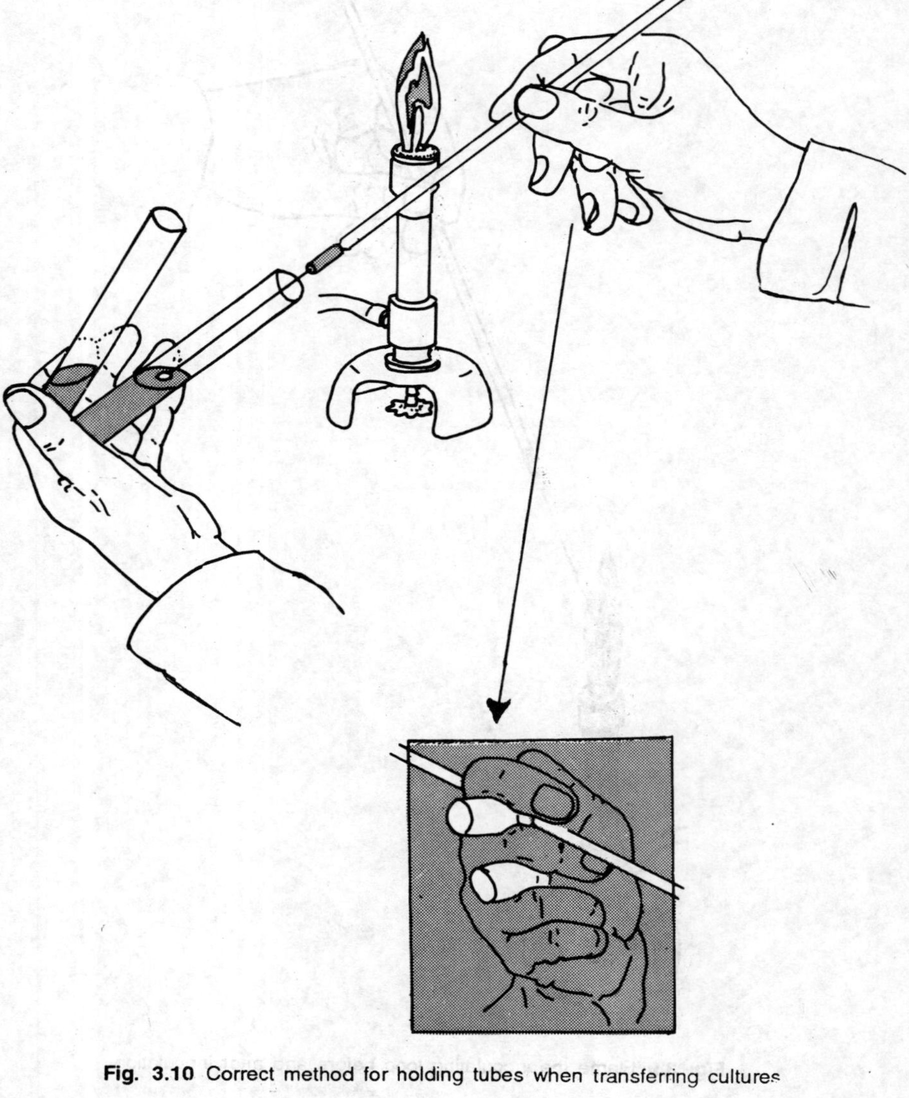

Fig. 3.10 Correct method for holding tubes when transferring cultures

Pure culture techniques

(1) *Streak plate :* It is a method of applying microbial culture on to agar surface with a loop or a bent glass rod (spreader). This is called streaking and a plate so prepared is streak plate (see Fig. 3.11).

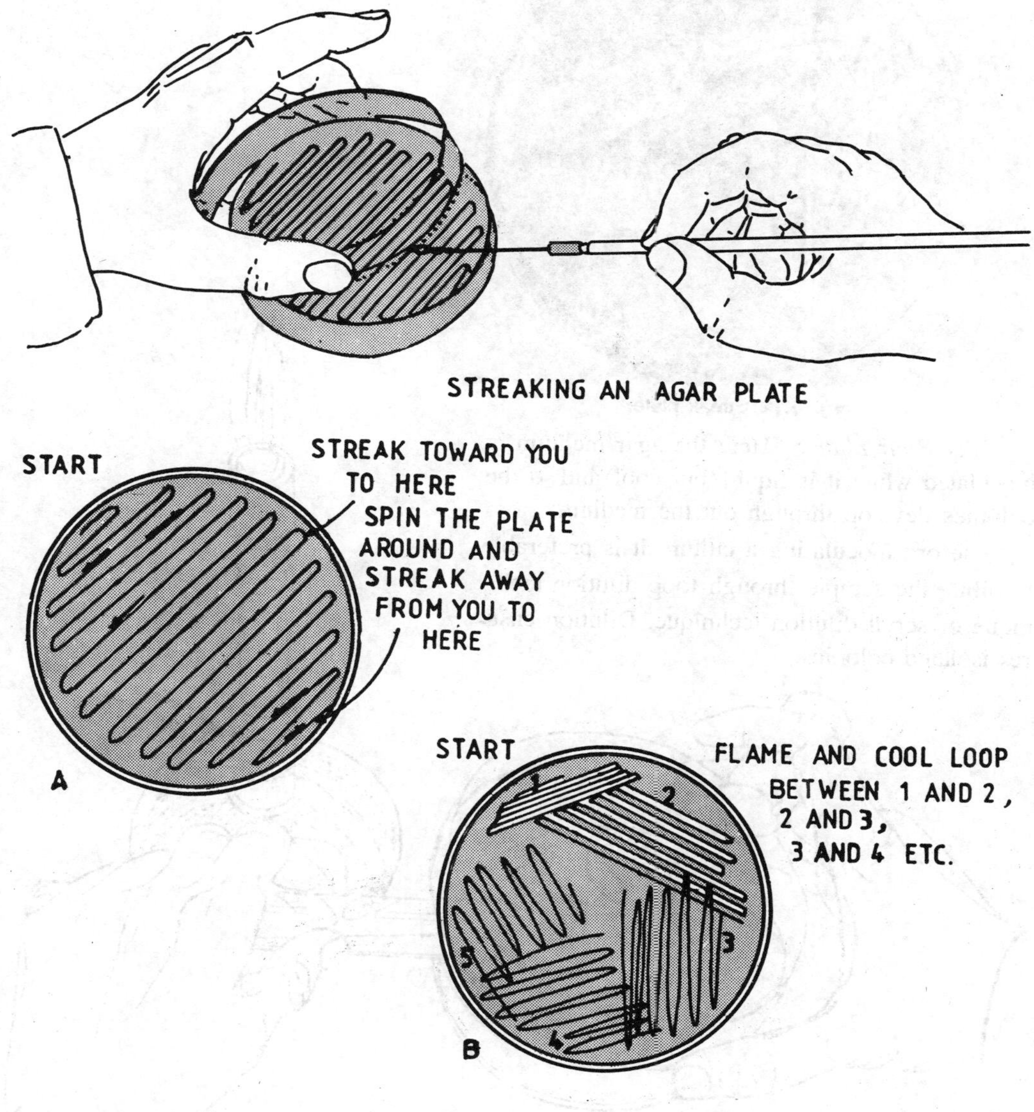

Fig. 3.11 Motion of loop in two methods of streaking plates.

During inoculation the closely packed cells of the start form colonies that run together. However, as the inoculum becomes thin towards the end, the colonies obtained here are isolated.

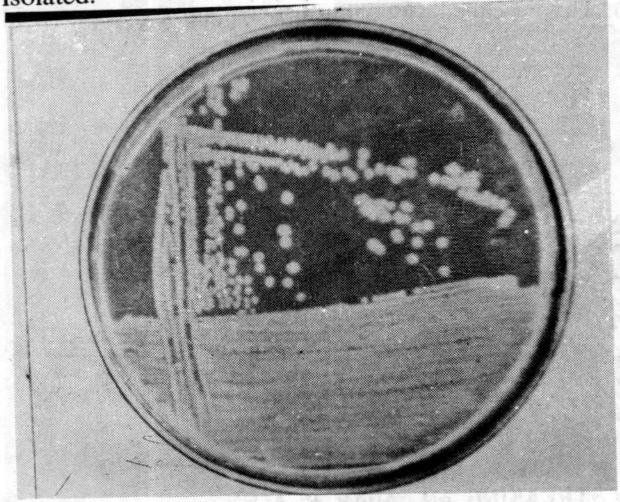

Fig. 3.12 Streak plate.

(2) *Pour plate :* Here the agar medium is inoculated while it is liquid (but cool) and so the colonies develop through out the medium.

Before inoculating a culture it is preferable to dilute the sample through loop-dilution technique or serial dilution technique. Dilution ensures isolated colonies.

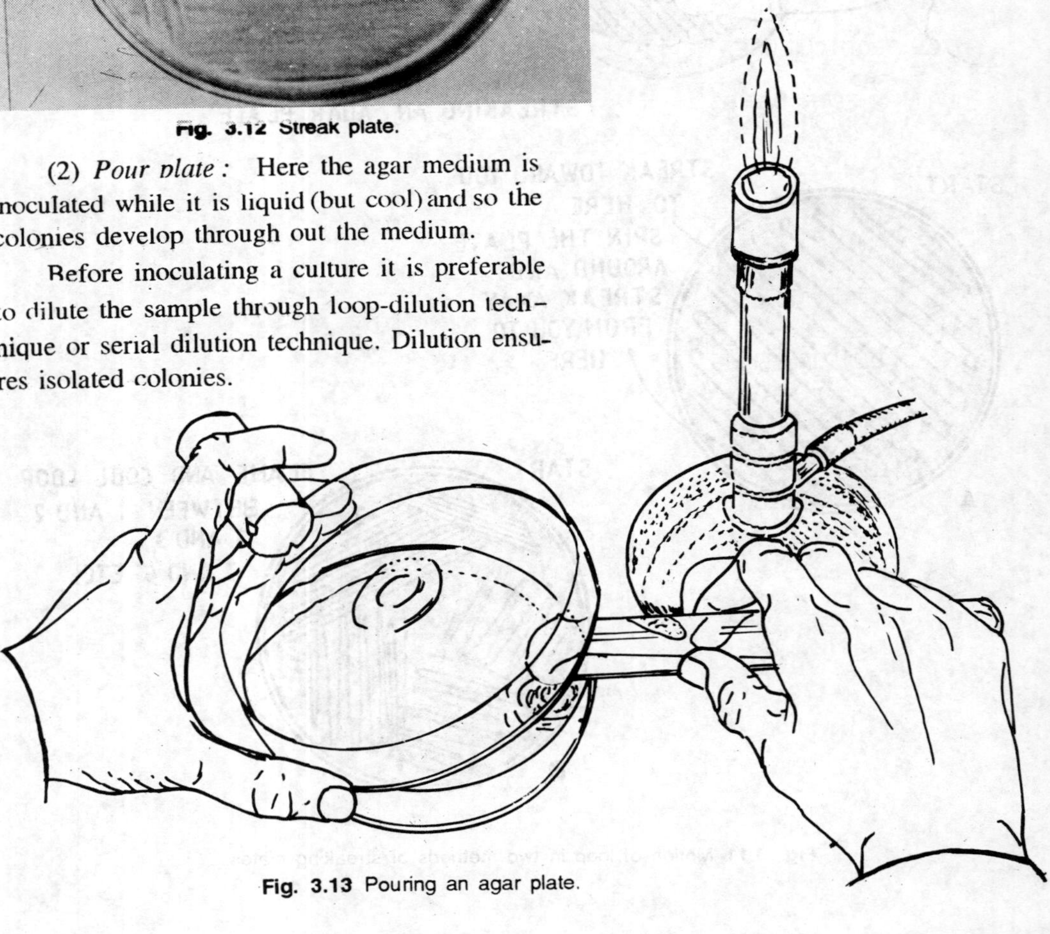

Fig. 3.13 Pouring an agar plate.

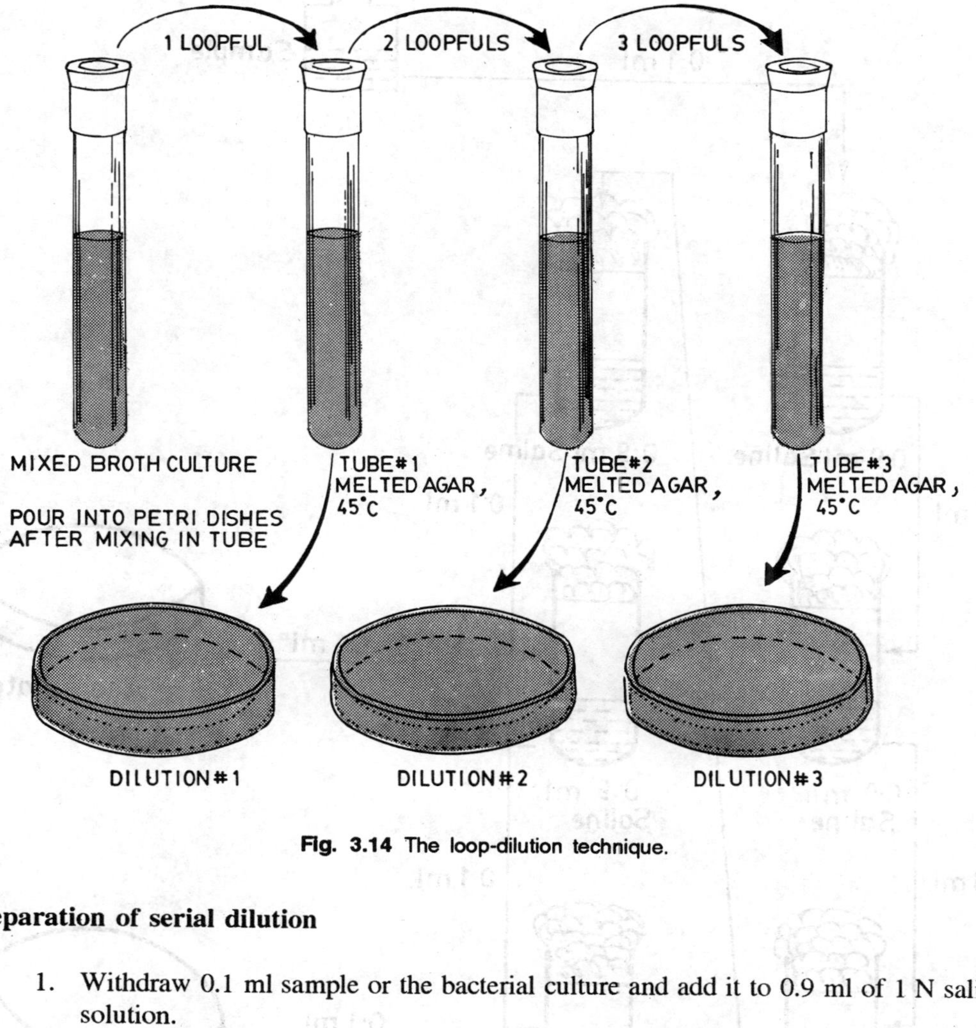

Fig. 3.14 The loop-dilution technique.

Preparation of serial dilution

1. Withdraw 0.1 ml sample or the bacterial culture and add it to 0.9 ml of 1 N saline solution.
2. Again take 0.1 ml of above inoculated saline and add to 0.9 ml of sterilized 1 N saline. This dilutes the bacterial sample to 10^{-2} times.
3. Again withdraw 0.1 ml of bacterial sample and add to 0.9 ml of saline.
4. This raises the dilution to 10^{-3} times.
5. In this way by repeated inoculation prepare dilution of 10^{-6}.
6. Now inoculate 0.1 ml of above sample to the plate by pour plate method.

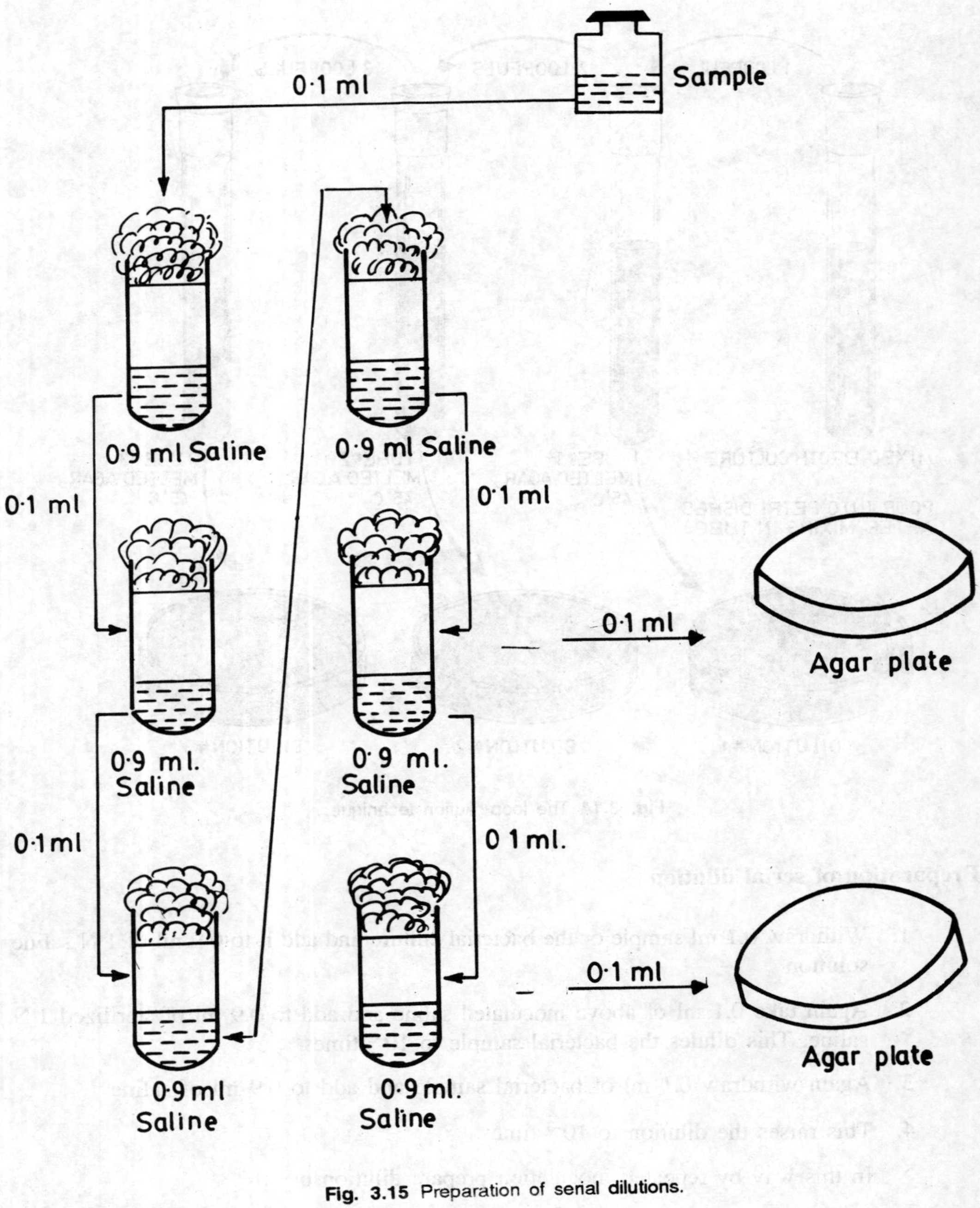

Fig. 3.15 Preparation of serial dilutions.

After sufficient incubation, the bacteria grow so as to form a mixed culture each having a characteristic colony morphology of its own. The nature of whole colony, the edge, the elevation and the surface fixture in itself provides maximum information about the type of bacterium. These characteristics are made use during identification of bacteria.

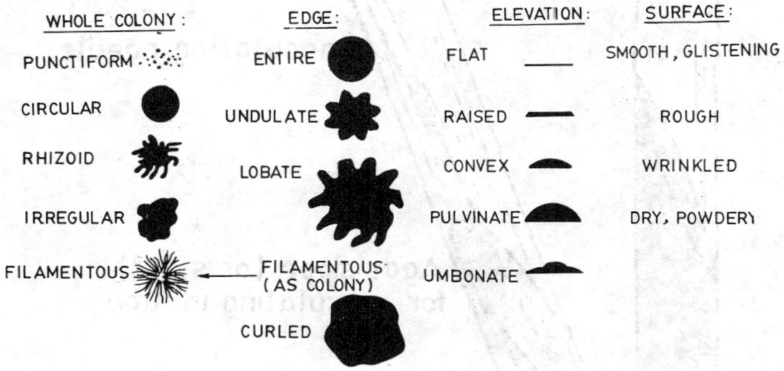

Fig. 3.16 Guide to terms used in colony descriptions.

Mixed to Pure culture

From a mixed culture, study the colony morphology and then transfer each of the colony type to different plates having the same isolation media as used before. Now watch if there is the development of only single colony type on the plate. If it is so the culture can be said to be pure culture. There are various methods to store a pure culture. However, the most common one which is followed generally is the transfer of pure culture from plate to an agar slant (a slant is a test tube 2/3rd filled with agar media and kept at an angle during solidification). The inoculated slant is kept in the incubator and after proper growth it is transferred to the refrigerator.

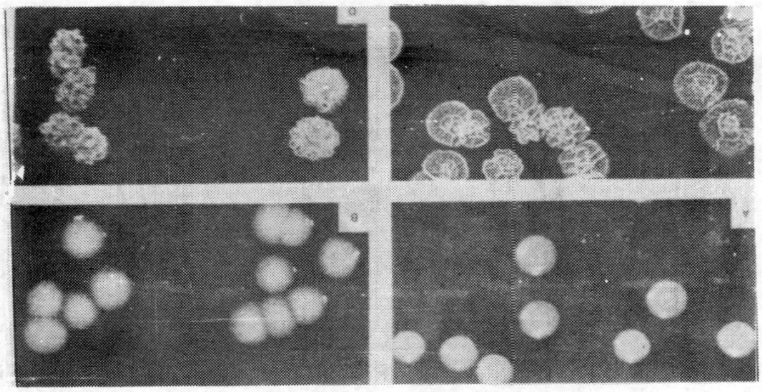

Fig. 3.17 Colony morphology.

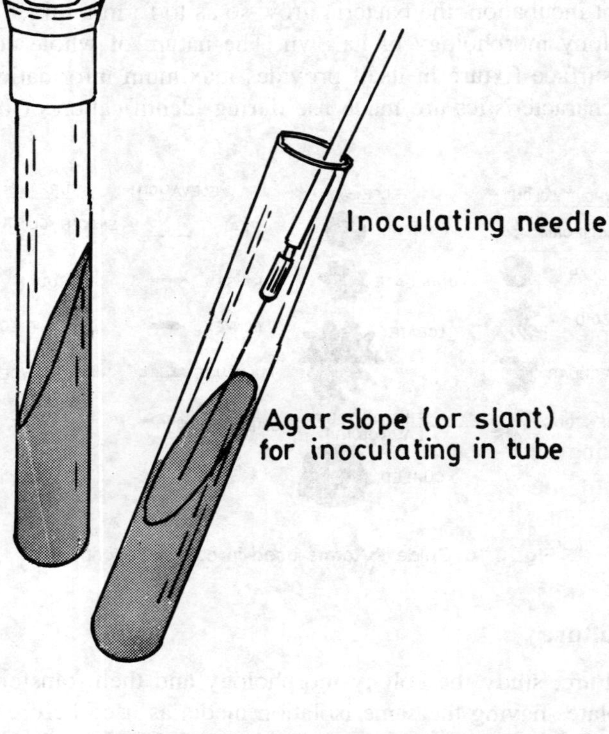

Fig. 3.18 Inoculation of the agar slopes.

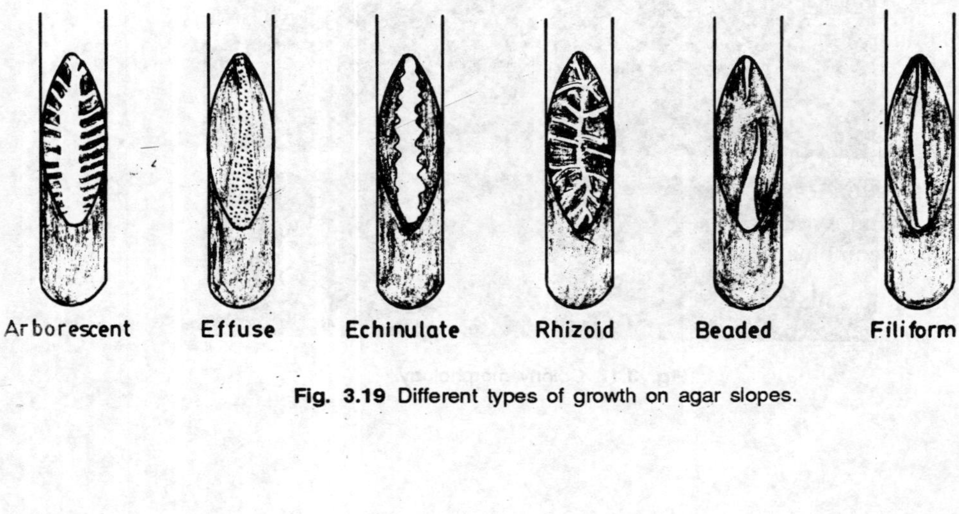

Fig. 3.19 Different types of growth on agar slopes.

4

Staining

OUTLINE

Hanging drop preparation • Staining • Simple stains • Differential stains • Structural stains

The study of bacterial morphology is performed in two ways :

1. Observing unstained cells live by hanging drop preparation.
2. Observing dead cells by making use of chemical nature of their unicellular body. This is achieved by staining.

Hanging Drop Preparation or Motility Test

1. Apply vaseline around the depression of the hanging-urop slide.
2. Using the inoculation loop, aseptically transfer one drop of the culture to the centre of a clean cover slip.
3. Invert the hanging-drop slide and centre its well over the drop of the culture. Press down on the edge of the cover slip so that the vaseline makes a firm seal.
4. Quickly and carefully turn the slide right side up so as to suspend the hanging drop in the well. Don't let the drop fall or touch the bottom of the well.
5. To examine, first locate its edge in centre of the microscopic field with low power objective and markedly lower the light. The edge will be visible as a bright wavy line against a dark background. Now the slide can be focussed under oil immersion.

Staining

It is a chemical or a physical union between the dye and like component of a cell. If it is a chemical reaction a new compound is formed and a simple washing with water does not liberate the bound dye but if purely physical it is easy to decolorise such stained organism. Usually it is a combination of chemical and physical reactions.

The main advantages of staining are that it

(i) provides contrast between microorganisms and their backgrounds, permitting differentiation among various morphological types;

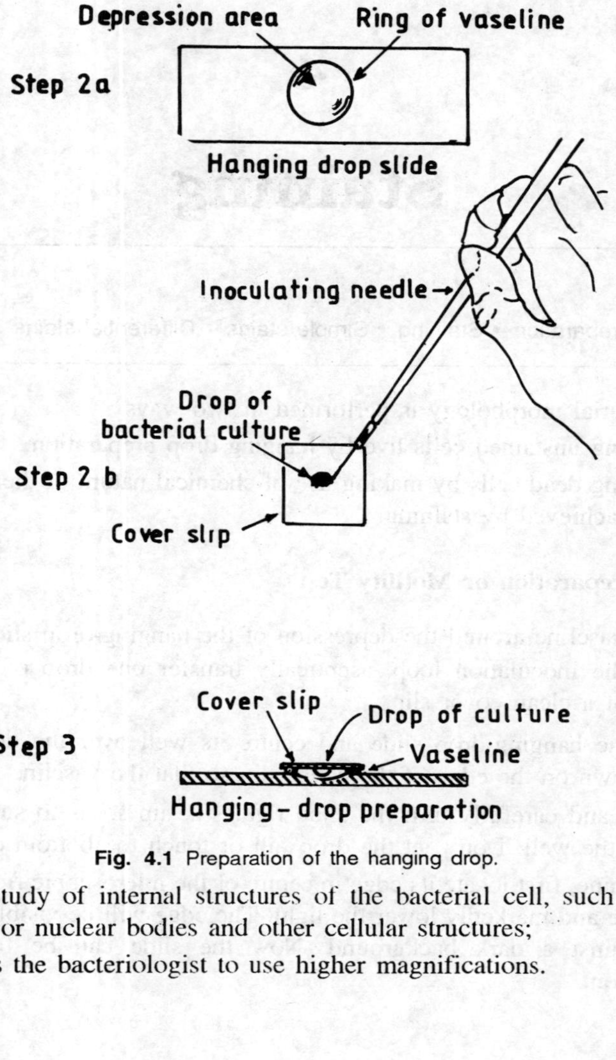

Fig. 4.1 Preparation of the hanging drop.

(ii) permits study of internal structures of the bacterial cell, such as the cell wall, vacuoles or nuclear bodies and other cellular structures;

(iii) it enables the bacteriologist to use higher magnifications.

Type of Stains

1. Simple stains
   (a) Acidic dyes
   (b) Basic dyes
   (c) Indifferent dyes

2. Differential stains
   (a) Gram stain
   (b) Acid fast stain

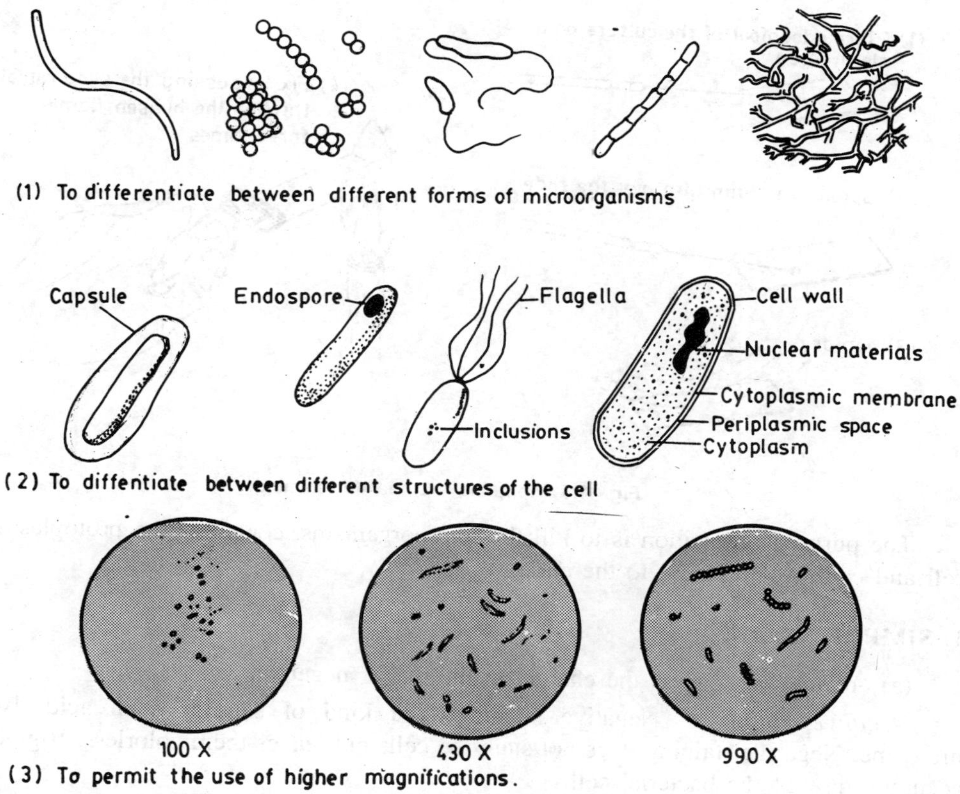

Fig. 4.2 Reasons for using stained preparations.

3. Structural stains
   (a) Endospore stain
   (b) Capsule stain
   (c) Flagella stain

Before staining it is essential to fix the bacterial sample on to the slide. Preparation of smear is done as follows :

(i) With a wire loop place a small drop of the broth culture or a loop full of bacteria on a clean slide.

(ii) Place a drop of water over it.

(iii) Spread the culture so as to form a thin film.

(iv) Allow slide to dry in the air or by holding it above a bunsen flame.

(v) Avoid excess heating.

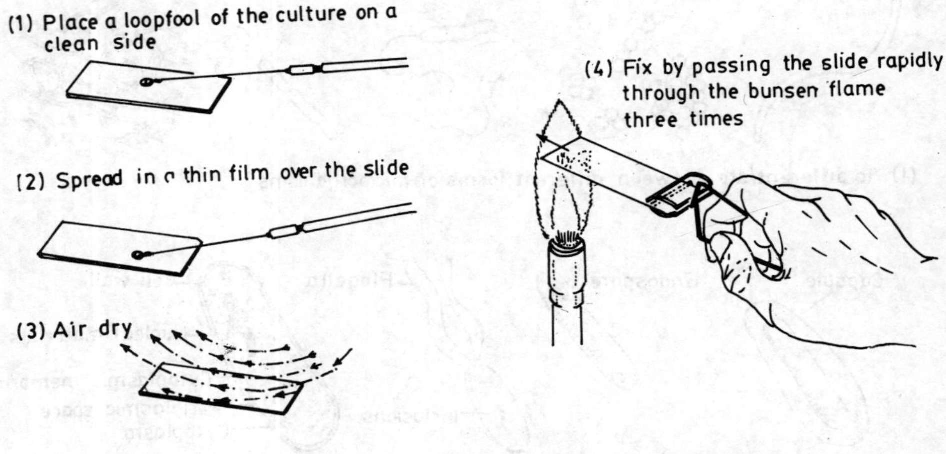

Fig. 4.3 Preparing a smear for staining.

The purpose of fixation is to kill the microorganisms, coagulate the protoplasm of the cell and cause it to adhere to the slide.

1. SIMPLE STAINS

(a) *Acidic dye :* Here the charge on the dye is negative.

Negative Staining : Negative staining is a kind of staining with acid dye, e.g., nigrosine. Negative staining does not stain the cells but leaves them colorless. It gives more accurate view of the bacterial cell.

(b) *Basic dye :* Here the dye carries a positive charge. For example, methylene blue.

MBC ⟶ Methylene blue + Chloride$^-$

(Bacterial cell –) (Na +) + (MB) + (Cl –) ⟶ (Bacterial cell –) (MB) + (Na + Cl –)

The bacterial cells develop a slight negative charge when the pH of the surrounding is near neutrality the negatively charged bacterial cell combines with positively charged methylene blue ion so that the cell is stained.

Crystal violet and carbol fuschin are some other examples.

(c) *Indifferent/Neutral Dye :* It is a complex salt of an acidic dye with a basic dye, e.g., Eosinate of methylene blue.

2. DIFFERENTIAL STAINS

(a) *Gram stain :* Developed by Christian Gram in 1884. By using this procedure, bacteria have been classified into 2 broad groups — gram-positive and gram-negative bacteria. Gram staining requires 4 different solutions

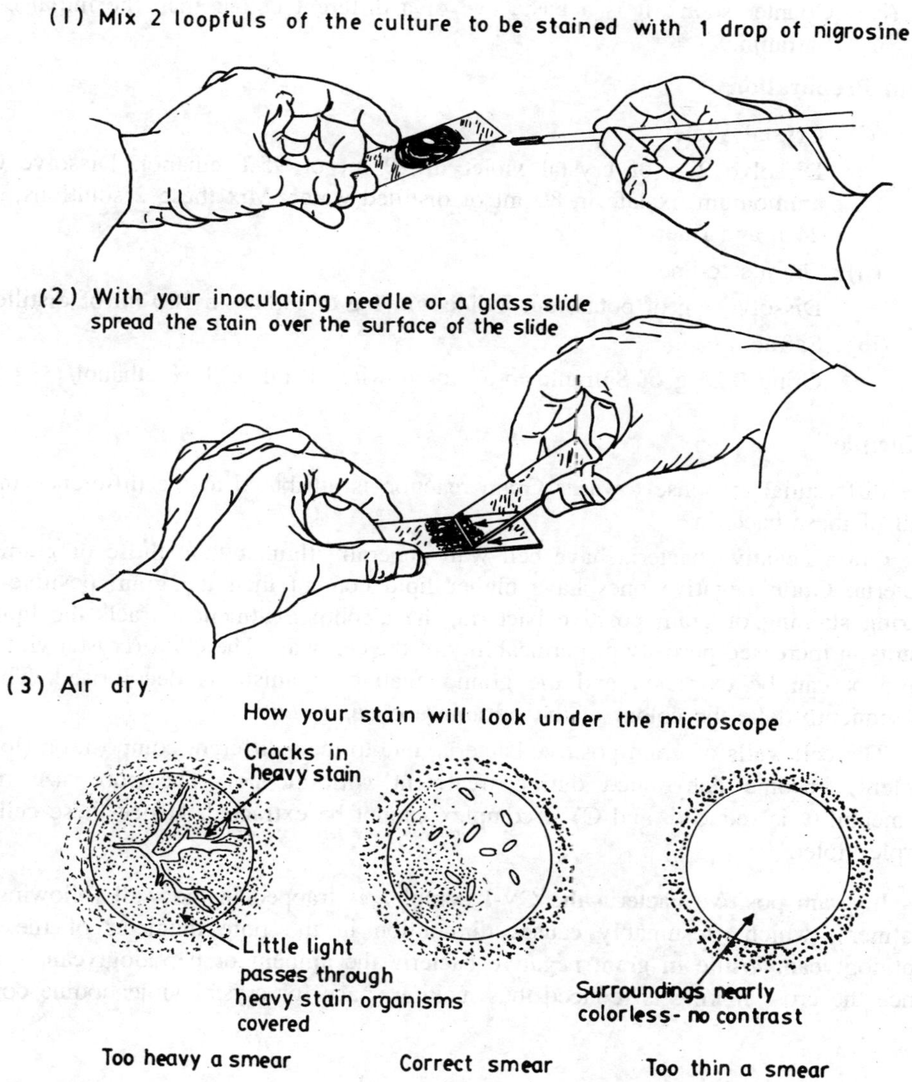

Fig. 4.4 The negative stain.

(i) A basic dye : Discussed previously.

(ii) A mordant : It increases the affinity or attraction between the cell and the dye, e.g., Iodine.

(iii) A decolorising agent : It removes the dye from a stained cell, e.g., Alcohol.

(iv) Counter stain : It is a basic dye of a different colour than the initial one, e.g., Safranin.

Stain Preparation

(i) Crystal Violet

Dissolve 2 g of crystal violet in 20 ml of 95% ethanol. Dissolve 0.8 g of ammonium oxalate in 80 ml of distilled water. Mix these 2 solutions, stand for 24 h and filter.

(ii) Gram's Iodine

Dissolve 2 g of potassium iodide and 1 g of iodine in 300 ml of distilled water.

(iii) Safranin

Grind 0.25 g of Safranin in a mortar with 10 ml of 95% ethanol.

Principle

The differential response towards Gram reaction is attributed to the difference in the cell wall of these bacteria.

Gram negative bacteria have cell wall generally thinner than those of gram positive bacteria. Gram negative ones have higher lipid content than the gram positive bacteria. During staining of gram positive bacteria, the alcohol treatment extracts the lipid, which results in increased porosity or permeability of the cell wall. The CV-I (crystal violet iodine) complex can be extracted and the gram negative organism is decolorised. These cells subsequently take the colour of the safranin a *counter stain.*

The cell walls of gram positive bacteria, due to their different composition (lower lipid content) become dehydrated during treatment with alcohol. The pore size decreases, permeability is reduced, and CV-I complex cannot be extracted. Hence these cells remain purple violet.

In gram positive bacteria the CV-I complex is trapped in the wall following ethanol treatment, which presumably causes diminution in the pore diameter of the walls of peptidoglycan. While in gram negative bacteria the amount of peptidoglycan is very low hence the cross linking is reduced thus making space for crystal violet iodine complex to escape.

Procedure

1. Prepare smear on a clean slide.
2. Stain with crystal violet for 30 seconds.
3. Rinse with water.
4. Flood the film with Grams iodine and allow it to act for 30 sec.
5. Rinse with water.

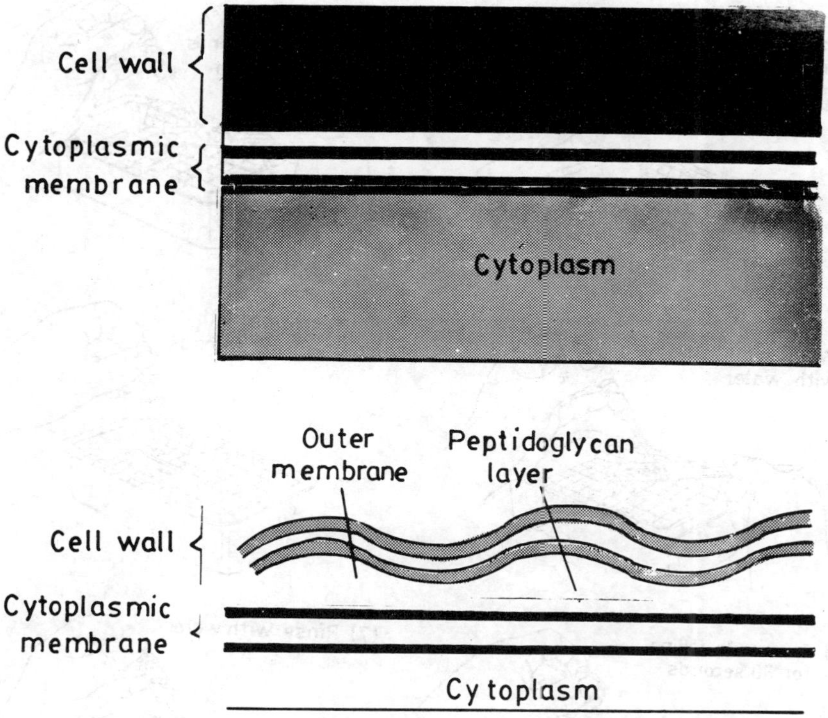

Fig. 4.5 Cell walls of gram positive and gram negative bacteria.

6. Decolorise with 95% alcohol.
7. Rinse with water.
8. Counter stain with safranin for 20-30 sec.
9. Rinse with water and blot dry.
10. Examine under oil immersion objective.

Table 5. Steps in Gram Stain

Step	Process	Results	
		Gram +	Gram –
Initial stain	CV for 30 sec.	Stains purple	Stains purple
Mordant	I_2 for 30 sec.	Remains purple	Remains purple
Decolorisation	95% alcohol for 10-20 sec.	Remains purple	Becomes colourless
Counterstain	Safranin for 20-30 sec.	Remains purple	Stains pink

Fig. 4.6 Gram staining procedure.

(b) *Acid fast stain :* The Ziel-Nehlson method employs carbol fuchsin, acid alcohol and a blue or green counter stain.

(i) Carbol fuchsin

1.	Basic fuchsin	58 g
2.	Crystalline phenol	25 g
3.	95% alcohol	50 ml
4.	D.W.	500 ml

Dissolve the fuchsin and phenol in alcohol over a warm water bath, then add water. Filter, before use.

(ii)	95% ethyl alcohol	970 ml
	Conc. HCl	30 ml

(iii) 0.5% methylene blue or malachite green, or 0.75% puric acid in distilled water.

Procedure

Pour carbol fuchsin on the slide and heat carefully until steam rises. Stain for 3-5 min. but do not allow to dry. Wash well with water. Decolorise using acid alcohol for 10-20 s. Changing twice and counter stain for 2 mins with methylene blue or malachite green.

Acid fast organisms, e.g., *Mycobacterium tuberculosis* and *Mycobacterium leprae.*

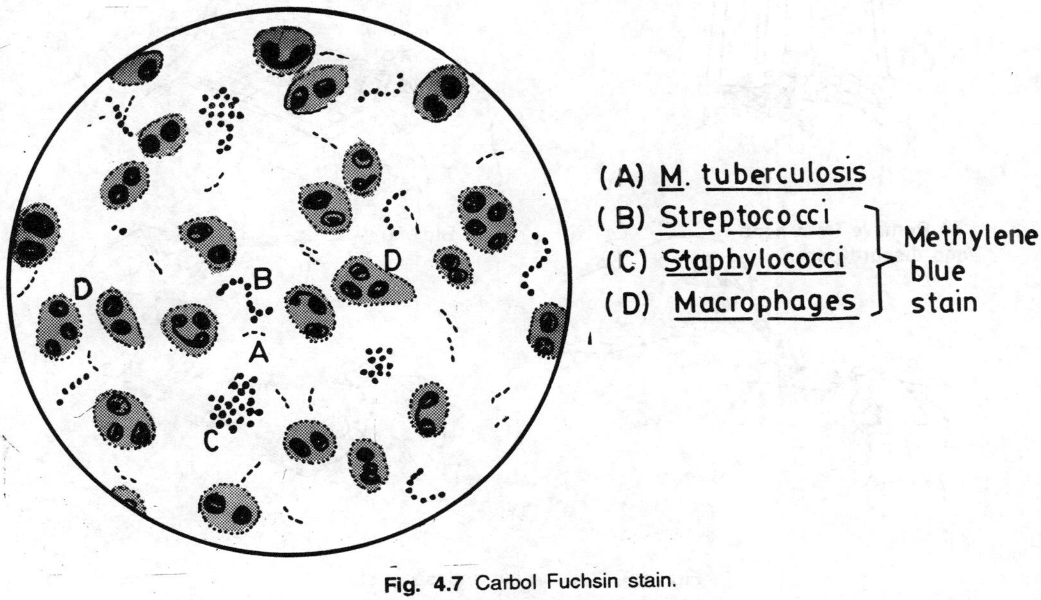

Fig. 4.7 Carbol Fuchsin stain.

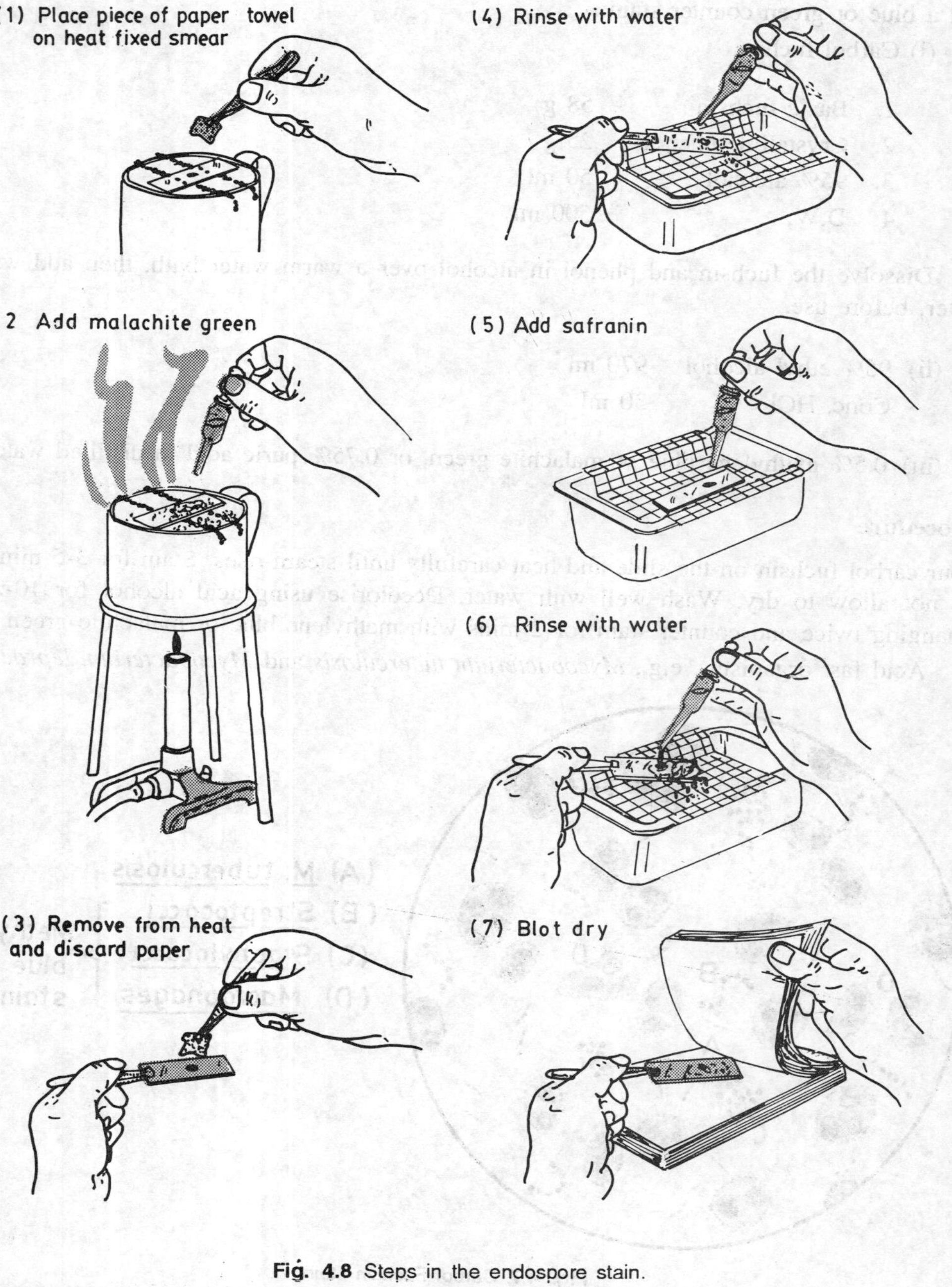

Fig. 4.8 Steps in the endospore stain.

3. STRUCTURAL STAINS

(a) *Endospore stain :* Endospores are highly resistant bodies, e.g., *Bacillus, Clostridium*, etc. They resist staining and, once stained they resist decolorisation and counter staining.

The malachite green (Schaeffer and Fulton) stain employs hot malachite green as an intense stain that is not removed from the endospore by washing and saffrain as a counterstain. Thus endospore stains green but rest of the cell or a cell with out endospore stains light red.

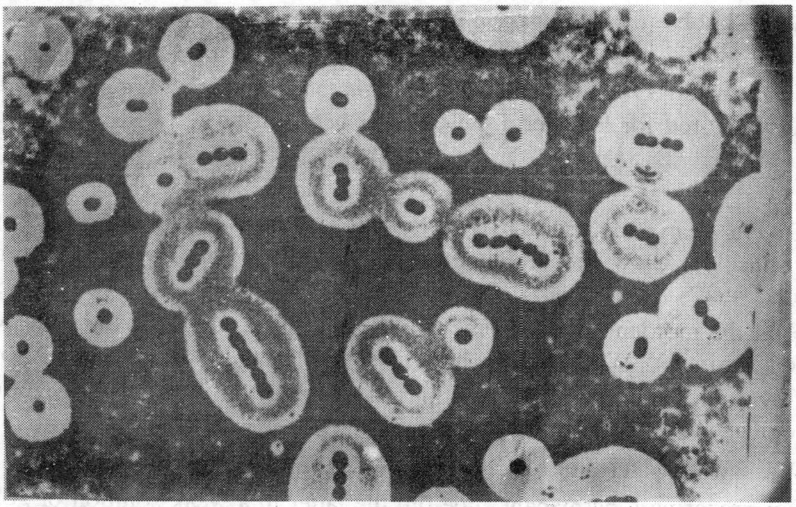

Fig. 4.9 Capsule stain.

Procedure

1. Prepare a thin smear on a clean slide.
2. Place the slide on staining rack above boiling water.
3. Cover the smear with small pieces of paper towel, keep saturated with malachite green (5% aqueous solution) and continue heating for 5 mins
4. Wash gently with water.
5. Counter stain with safranin for 30 secs.
6. Wash with water and blot dry.
7. Examine under oil immersion objective.

(b) *Capsule stain :* Place a small drop of Indian ink on a clean slide. Mix into it a small loopful of bacterial culture. Drop a cover glass and blot off excess ink. Examine.

For dry preparations mix one loopful of Indian ink with one loopful of suspension of organism in 5% dextrose solutions at one end of a slide. Allow to dry and pour a few drops of methyl alcohol on it. Keep the slide over the flame to fix. Stain for a few sec. with 0.5% aq. solution of methyl violet. The capsule will appear as haloes in blue cell of bacterium under microscope.

(c) *Flagella stain :*

(i) Mordant — 20% tannic acid solution	3 parts
5% tartar emetic solution	2 parts
D.W.	5 parts

Mix and redissolve the heavy precipitate by boiling.

(ii) Silver solution : Dilute a saturated solution of silver sulphate with an equal volume of distilled water. Add 33% monoethylamine till the precipitate that is formed is redissolved. Solution can be stored for long term, heat gently before using.

Procedure

1. Take a clean slide and transfer a loopful of bacterial suspension holding the slide vertically. Allow the drop to run to the other end. Dry the film in air or in incubator.
2. Heat the mordant in a test tube to a boiling point and flood the slide with it wait for 2 minutes. Wash well in hot water and finally in cold water. Allow to dry.
3. Cover the film with silver solution till thick portion of the film becomes dark to brown and metallic sheen appears on the edge.
4. Wash with water and blot dry.
5. For preparing a permanent slide dip the latter in a weak solution of gold chloride for 1 or 2 hr. Wash and mount in canada balasam.

5

Biochemical Tests

OUTLINE

• Amylase activity • Caseinase activity • Carbohydrate digestion • Catalase tes⁺ • Urease test • H_2S production • IMViC reactions • Litmus milk test • Nitrate reduction • Oxidase test • Pectinase test • Gelatinase test • Lipase test • Gluconate test

1. AMYLASE ACTIVITY

Amylase is an enzyme confined to the starch hydrolysing organisms. It can be tested using Lugol's iodine.

Principle

Starch is made up of two polysaccharides :

(i) Amylose, soluble in water and a polymer of 200-1000-D-glucose molecules limited by 1-4 glycosidic linkages. It dissolves in hot water giving blue colour with iodine.

(ii) Amylopectin is insoluble in hot water and is a polymer of 100 chains of D-glucose molecules united by 1-4 glucosidic linkages. The chains are united by 1-6 glycosidic linkage.

Mode of Action of Amylase : It is a complex of α and β amylases. Both degrade amylopectin to maltose unit, the latter is acted upon by β-glucosidase to give glucose. While α-amylase alone degrades amylose to yield monomer glucose.

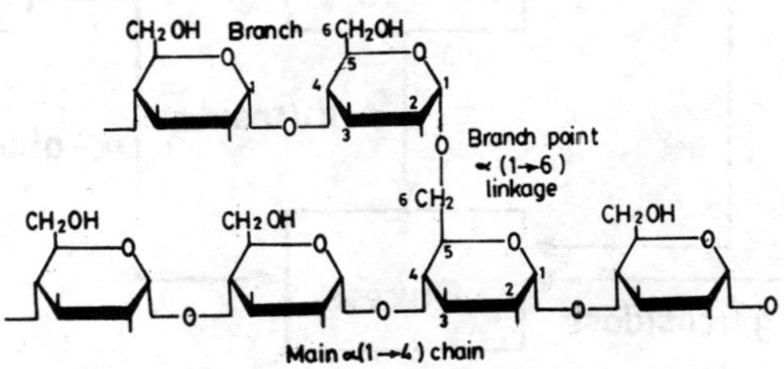

Fig. 5.1 An α (1 → 6) branch point in amylopectin.

Procedure

Reagent and media preparation.

A. Starch agar

1. Prepare a 10% solution of soluble starch in water and steam for 1 h.
2. Add 20 ml of this solution to 100 ml of melted nutrient agar and pour plates.

Composition of nutrient agar

Beef extract	3 g
Peptone	5 g
NaCl	5 g
$NaNO_3$	5 g
DW	1000 ml
pH	7.0

B. Lugol's Iodine

Iodine	1 g
Potassium iodide	2 g
D.W.	300 ml

Inoculate starch agar and incubate for 24 hrs. Flood the plate with Lugol's iodine and observe for the clear zones.

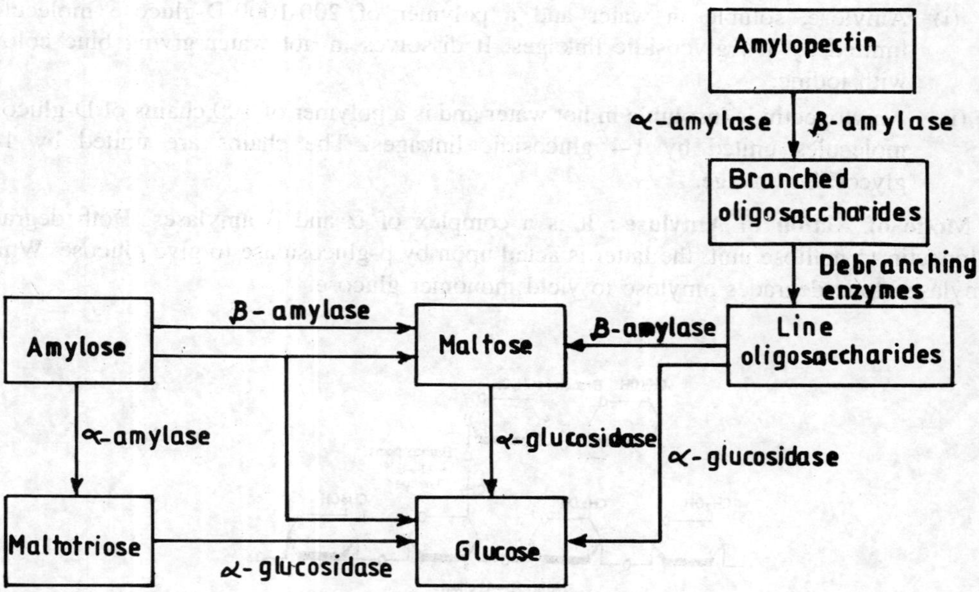

Fig. 5.2 Enzymatic conversion of starch to glucose initiated by α- and β-amylases.

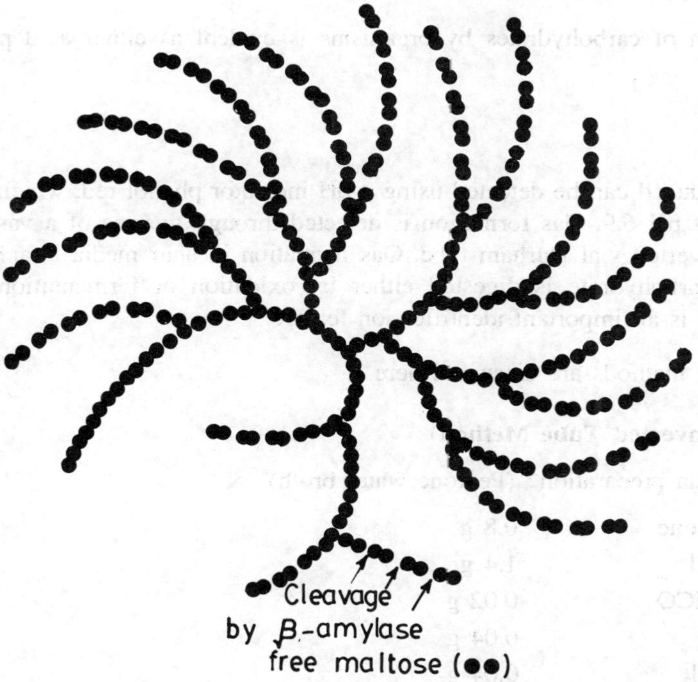

Fig. 5.3 Action of β-amylase on amylopectin.

2. CASEINASE

Casein is the principal protein of milk. It exists as a colloidal suspension that gives milk its opaque whiteness. Many bacteria are equipped with caseinase enzyme that hydrolyses this protein. This reaction of peptonisation is useful for identification of microorganisms.

Procedure

Media Preparation

1. Reconstitute milk powder and mix equal volumes of this and double strength nutrient agar melted at 55°C mix and pour plates.
2. Inoculate the plates with a loop to place cells at the centre only.
3. Incubate at 37°C for 24-48 hrs.
4. Colonies of organisms that produce caseinase will be surrounded by clear zone. Areas where caseinase has not been produced will remain opaque.

3. CARBOHYDRATE DIGESTION

The digestion of carbohydrates by organisms is evident as either acid production or gas production.

Principle

The acid produced can be detected using a pH indicator phenol red, which is red at pH 8.5 and yellow at pH 6.9. Gas formation is detected through the use of a vasper (or vaseline) seal or an inverted vial Durham tube. Gas formation in agar media is achieved by bubble formation. Carbohydrate is digested either by oxidation or fermentation which in gram negative rods is an important identification feature.

3 simple methods are discussed here.

Method I (Inverted Tube Method)

(A) Media preparation: (Peptone water broth)

Peptone	0.8 g
NaCl	1.4 g
$NaHCO_3$	0.02 g
KCl	0.04 g
$CaCl_2$	0.04 g
KH_2PO_4	0.24 g
$Na_2HPO_4 \cdot 2H_2O$	0.88 g
D.W.	1000 ml

Steam the ingredients at 100°C for 20 min, and filter through double fold Whatman No. 1 filter paper. The pH is adjusted to 7.2. Autoclave.

Addition of Sugar : Prepare 1% solution of the desired sugar. Autoclave at low pressure and add to above solution in order to make to a final concentration of 10% by volume.

Addition of Indicator (phenol red) : Dissolve 0.1g of phenol sulphonaphthalein in 14.1 ml of 0.02 N NaOH and dilute to 250 ml with water, autoclave and add to basal medium making a final concentration of 0.4% by volume.

Phenol red is yellow at pH 6.8 (acid production) and red at pH 8.4 (no acid).

(i) Transfer media in test tubes and invert a Durham tube in each of these tubes.
(ii) Inoculate each tube with the culture to be identified.
(iii) Incubate at 37°C.
(iv) Observe the change in colour, if any.

Method II (Disc Method)

It differs from the above method in that the sugar is not added to medium directly but sterilised filter paper discs are impregnated with the sugar solution and then placed on the inoculated agar plates aseptically. Observe after 48 hrs.

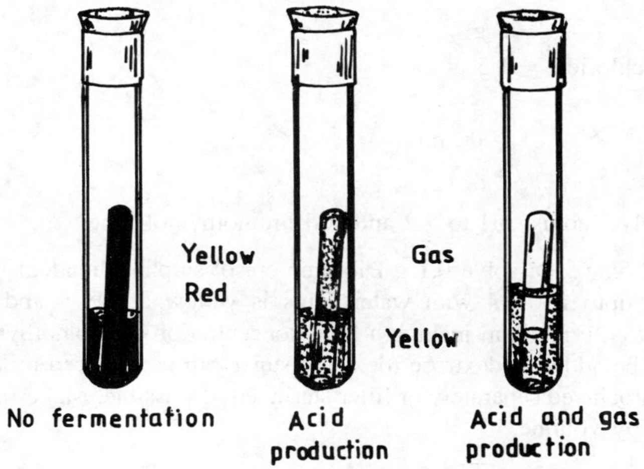

Fig. 5.4 Reaction in phenol red fermentation tubes.

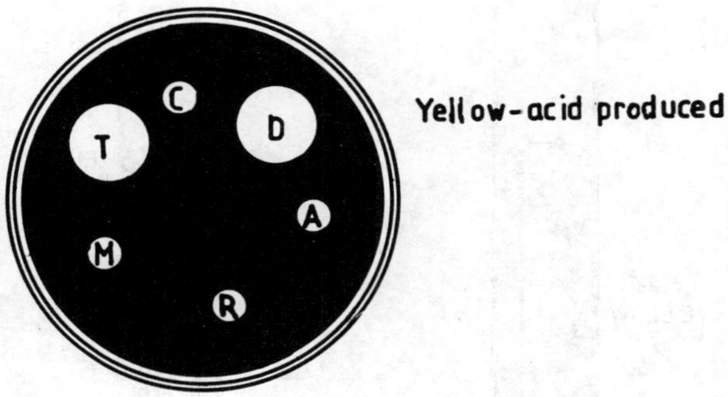

Fig. 5.5 The disc method for determining carbohydrate fermentation. Letters indicate type of carbohydrate.

Method III

It illustrates whether the carbohydrate is utilised oxidatively or fermentatively. Here oxidation fermentation medium is used (OF) with bromothymol blue as a pH indicator.

Tubes are inocuiated by stabbing. One tube is incubated aerobically while other is kept in anaerobic condition. Fermented cultures will turn both tubes yellow while oxidative culture will turn only one tube yellow. In some oxidative culture tubes no acid end product will be detected. If the colour production is homogeneous, it also confirms the motility of the organism.

OF Agar medium / OXIDATION-FERMENTATION MEDIUM

Peptone	2 g
Sodium chloride	1.5 g
K_2HPO_4	1 g
D.W.	1000 ml
Agar	3 g

Heat to dissolve, adjust pH to 7.2 and add bromothymol blue.

Bromothymol blue : Dissolve 0.1 g Dibromo-cresol sulphnaphthalein in 8 ml of 0.02 N NaOH and make upto 250 ml with water. This is yellow at pH 6 and blue at pH 7.6. Sterilize and add to OF medium making a final concentration of bromothymol blue solution as 0.4%, sugar to be added is dextrose (dextrose sugar can be replaced by any other desired sugar) which is autoclaved separately or filter sterilized, if possible. Make final concentration of sugar as 0.1% by volume.

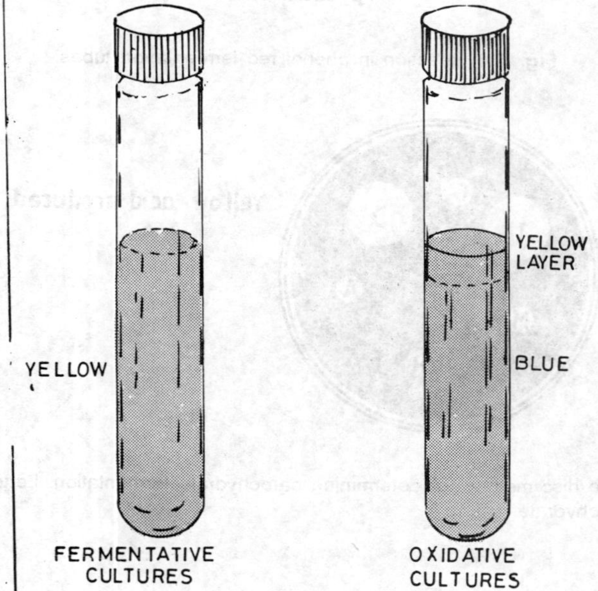

Fig. 5.6 Oxidation fermentation tubes.

Procedure

1. Inoculate the slants of OF medium by stabbing once.
2. Cover one tube with a thin layer of vaspar. (Vaseline 1 lb and paraffin 1 lb. Melt and mix.)
3. Incubate for 48 hrs at 37°C.
4. Observe for production of acid and gas from dextrose. Fermentative cultures turn both tubes yellow, while oxidative cultures will turn only the open tube yellow.

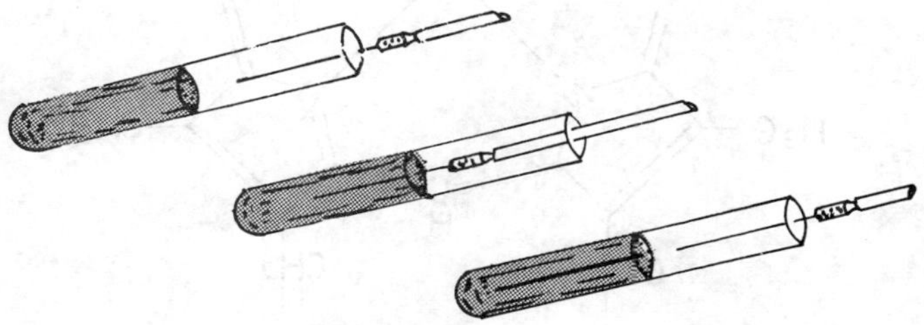

Fig. 5.7 Preparing an agar-stab culture.

4. CATALASE ACTIVITY

Catalase, an enzyme catalyses the breakdown of hydrogen peroxide to release free oxygen.

$$2H_2O_2 \xrightarrow{\text{Catalase}} 2H_2O + O_2$$

The enzyme contains the hemiporphyrin structure (Fig. 5.8). This porphyrin ring structure is characteristic of not only catalase but also of cytochromes.

Method I

Add a few drops of 3% H_2O_2 to the slope and broth cultures and observe closely for the appearance of oxygen bubbles and a surface-broth accumulation.

Method II

Dip a glass capillary in the beaker containing 3% H_2O_2 and allow capillary action to fill the tube to a height of about 20 mm. Touch the capillary on one of the colonies. Observe the bubbles in capillary for positive catalase reaction, e.g., *Lactobacillus*.

5. UREASE

The hydrolysis of urea by urease releases NH_3

$$\underset{\text{Urea}}{(NH_2)_2C=O} + H_2O \xrightarrow{\text{Urease}} 2NH_3 + CO_2\uparrow$$

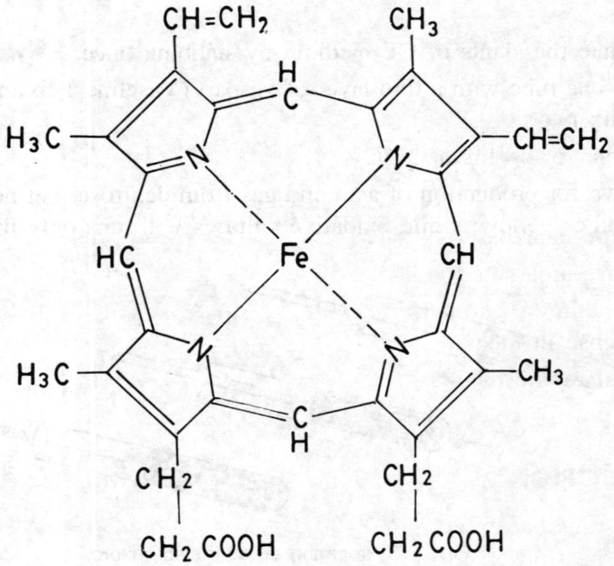

Fig. 5.8 Heme, an example of an iron chelate porphyrin.

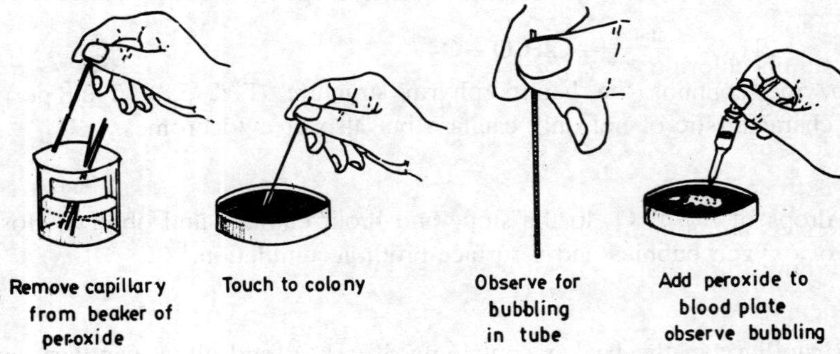

Fig. 5.9 Catalase activity.

Principle

Since the NH_3 production raises pH of the medium, urease activity can be detected by the change of the phenol red indicator to purple.

Urea agar

Peptone	1 g
Glucose	1 g

NaCl	5 g
K_2HPO_4	5 g
Phenol red	6 ml of 1 : 500 aqueous solution
Agar	20 g
DW	1000 ml
pH	6.8-6.9

1. Prepare agar base and sterilize.
2. Prepare urea solution 40% separately. Sterilize using bacteriological filter.
3. Add solution 2 to agar base so as to make a final concentration of 20% of urea. Dispense in tubes to prepare slopes.
4. Inoculate and incubate.
5. Observe for colour change of phenol red indicator to purple. e.g. *Proteus.*

6. H_2S PRODUCTION

The activity of some bacteria on S-containing amino acids frequently results in the liberation of H_2S. This is best observed in a triple sugar iron agar slant.

T.S.I. Agar

Trypticase	10 g
Peptone	10 g
Sodium chloride	5 g
Lactose	10 g
Sucrose	10 g
Dextrose	1 g
Ferrous ammonium sulphate	0.2 g
Sodium thiosulfate	0.2 g
Phenol red	0.025 g
Agar	13 g
D.W.	1000 ml
pH	7.3

$$\underset{\text{Cysteine}}{H_2N{-}\overset{\overset{\displaystyle CH_2{-}SH}{|}}{C}H{-}COOH} + 2H \longrightarrow \underset{\text{Alanine}}{H_2N{-}\overset{\overset{\displaystyle CH_3}{|}}{C}H{-}COOH} + H_2S$$

Table 6 Key Reaction For TSI Agar

	REACTIONS IN BUTTS AND SLANTS	SUGAR FERMENTATION	POSSIBLE ORGANISMS
1.	A and G in butt; A on slant H_2S not produced °1	Glucose – AG Lactose and/or Sucrose – AG	*Escherichia, Klebsiella,* intermediate coliform bacteria; *Proteus, Paracolons*
2.	A and G in butt Alk. on slant : H_2S produced	Glucose – AG Lactose and Sucrose – not fermented	*Salmonella, Proteus, Paracolons*
3.	A – only in butt Alk. on slant : H_2S not produced °2	Glucose – AG Lactose and Sucrose – not fermented	*Salmonella, Shigella, Paracolons*
4.	Alk. or neutral on butt Alk. on slant : H_2S not produced	None	*Alcaligenes, Pseudomonas*
5.	No change in butt in 24 hours A on butt in 48 hours A on slant : H_2S produced	Glucose and Sucrose – slight A; Lactose – not fermented	*Pasteurella* °3

A = Acid (yellow). Alk. = Alkaline (red). G = Gas. °1 = *Cit. freundii* produced H_2S ; °2 = *Salm. typhi* produced a small amount of H_2S in the butt; and occasional strain may be encountered to give a small amount of gas in the butt; °3 = P. multocida is rarely encountered. Enterococci may give a similar type of reaction.

7. IMVIC REACTIONS

Indole is a nitrogenous compound and a degradation product of amino acid tryptophan of various bacteria.

Indole can be detected as follows

(a) Put a strip of filter paper (soaked in a saturated solution of oxalic acid and dried) inside the overnight grown culture tube. Keep the filter paper suspended by the cotton plug. The paper becomes pink, when the test is positive.

(b) Kovac's method

Kovac's Reagent

p-dimethyl amino benzaldehyde	1 g
Iso-amyl alcohol	15 ml

To an overnight peptone culture, add Kovac's reagent to layer it, shake and allow the reagent to float. Indole is detected by the appearance of a pink colour at the top.

Peptone water culture is prepared in the same way as in Method I of carbohydrate digestion.

Methyl-Red Test (MR Test)

It detects organisms that do not convert acidic products to neutral products and produce a final pH lower than that of organisms producing neutral products. Because of the lower pH, the methyl-red indicator changes to a red colour, which is a positive test.

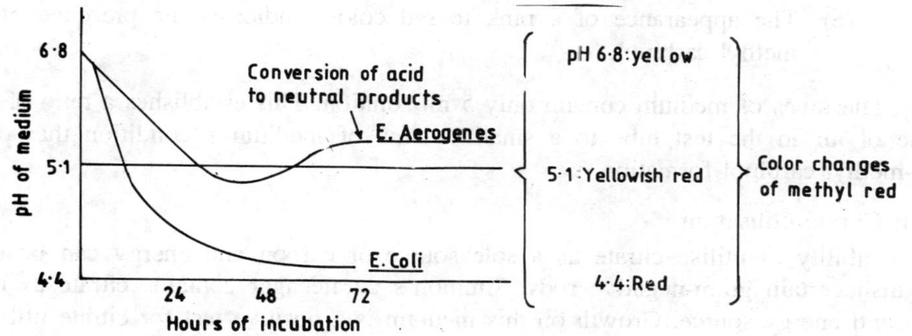

Fig. 5.10 Relation between acid and acetylmethylcarbinol production by *Enterobacter aerogenes* and *Escherichia coli* as a function of time.

Composition of glucose-phosphate broth

Glucose	5 g
Proteose peptone	5 g
K_2HPO_4	5 g
D.W.	1000 ml

1. Inoculate a tube of glucose phosphate broth with unknown culture.
2. Incubate for 48 hrs at 37°C.
3. Test for acid production by adding a few drops of an alcoholic solution of methyl red to the rest of the culture. A distinct red colour is considered to be a positive test; yellow is negative.

Methyl Red : 0.1 g of methyl red in 300 ml alcohol. Dissolve and make volume upto 500 ml.

(c) Voges-Proskauer Test (VP Test)

In some bacteria, e.g., *Bacillus* instead of accumulating mostly, acidic products from the fermentation of glucose, it is converted to metabolic intermediate, pyruvic acid and then to neutral products (acetylmethyl carbinol – $CH_3COCHOHCH_3$) plus CO_2.

Procedure

1. Inoculate a tube of glucose-phosphate broth with the unknown culture.
2. Incubate for 48 hrs at 37°C.
3. Test for acetyl methyl carbinol :
   (a) Decant about one-quarter of the culture into a clean test tube.
   (b) Add 0.5 ml (8-10 drops) of the α-naphthol solution (5% solution in alcohol).
   (c) Add 0.5 ml (8-10 drops) of the 40% KOH solution containing 0.3% creatine.
   (d) Shake thoroughly and allow to stand for 5-30 minutes.
   (e) The appearance of a pink to red colour indicates the presence of acetyl methyl carbinol.

Note : The tubes of medium contain only 5 ml solution. This establishes a ratio of a large volume of air in the test tube to a small volume of medium a condition that promote acetyl-methyl carbinol formation.

(d) Citrate utilisation

The ability to utilise citrate as a sole source of carbon and energy can be used to distinguish certain gram-negative rods. Simmon's citrate agar contains citrate as its only carbon and energy source. Growth on this medium is a positive test for citrate utilisation.

Simmon's Citrate Agar

$MgSO_4$	0.2 g
$(NH_4)_2HPO_4$	1.0 g
K_2HPO_4	1.0 g
Sodium citrate	2.0 g
NaCl	5.0 g
Agar	15 g
Bromothymol blue	0.08 g
D.W.	1000 ml

Prepare slants and inoculate.

8. LITMUS-MILK REACTION

Here litmus powder is added to detect the reaction after fermentation of lactose, when it becomes red or may show clotting or peptonisation of milk.

Preparation of litmus milk media.

(1) Dissolve 150 g of skimmed milk powder to a litre of water. Add sufficient amount of 2.5 percent aqueous solution of azolitmin.

(2) Inoculate and observe the change in the reaction of milk it will turn red if acid is produced and blue if alkali is produced.

(3) Clotting is followed by peptonisation of curd.

9. NITRATE REDUCTION

The reduction of nitrate is brought about by several soil inhabiting bacteria — nitrate, ammonia, nitrous oxide, nitrogen gas etc. are end products of nitrate reduction. They serve as H_2-acceptors for these species in absence of oxygen.

Procedure

(1) Nitrate broth

Nutrient broth plus 5 g/lit KNO_3 or $NaNO_3$

Nutrient Broth

Beef extract	3 g
Peptone	5 g
NaCl	5 g
$NaNO_3$	5 g
D.W.	1000 ml
pH	7.0

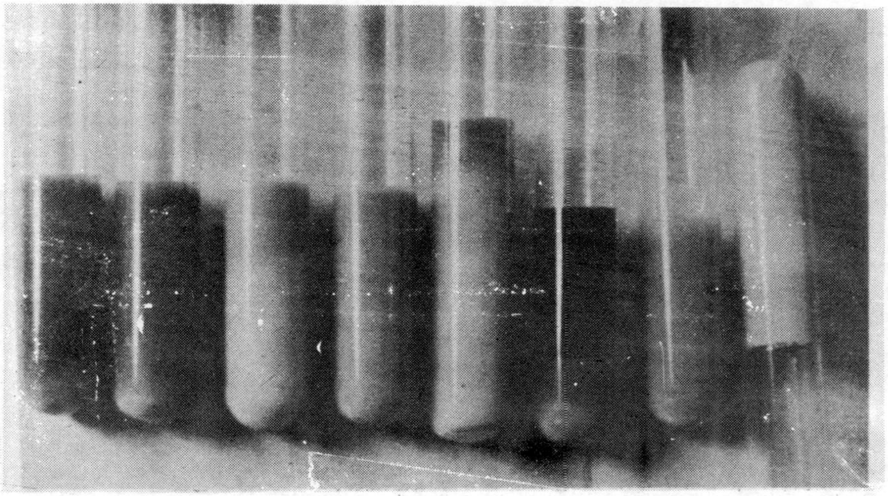

Fig. 5.11 Litmus milk reactions.

1. Inoculate one tube of nitrate broth having a Durham tube with unknown culture.
2. Incubate at 37°C for 24 h.
3. Observe for gas formation if there is no gas incubate further.

4. Add 1 ml of α-naphthylamine reagent and 1 ml of sulfanilamide reagent to the culture tubes after incubation.
5. A red colour with in 80 seconds shows nitrate reduction.

Reagent Preparation

(a) Sulfanilamide : Sulfanillic acid (0.8%) dissolves 8 g of reagent in 1000 ml of 5 N acetic acid, store in a glass stoppered bottle.

(b) α-Napthylamine (0.5%) : Dissolve 5 g of reagent in less than 1000 ml of 5 N acetic acid (30%) by heating gently. Filter with through washed adsorbent cotton. Store in a glass stoppered brown bottle in a refrigerator.

$$NO_3^- \longrightarrow NO_2^- \longrightarrow NH_3 \longrightarrow N_2O$$

10. OXIDASE TEST

It measures the ability of a microbe to oxidise aromatic amines, e.g., p-amino methyl aniline to yield colored end products. Oxidation on one hand correlates with high cytochrome oxidase activity of some bacteria e.g. *Pseudomonas* and on the other a negative oxidase reaction shows presence of the enteric bacteria. e.g. *Escherichia coli.*

Procedure

Reagent : 1% solution of tetramethyl-p-phenylene diamine hydrochloride.

(1) Using a sterilised loop, pick growth from a well-isolated colony and rub into area 1 of the test strip (a filter paper disc, impregnated with oxidase reagent).

(2) Wait for 30 seconds and observe the inoculated area.

(3) A blue or deep purple colour indicates a positive test.

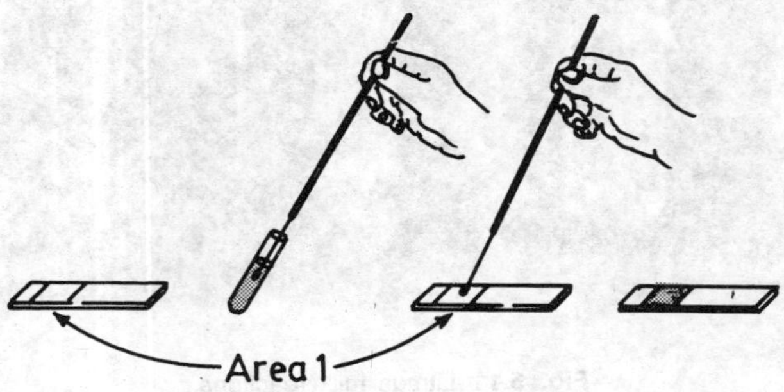

Fig. 5.12 Oxidase test.

(4) Instead of using paper strips, solution can directly be poured on to the agar plate. However, a test strip keeps the reagent stable for longer time.

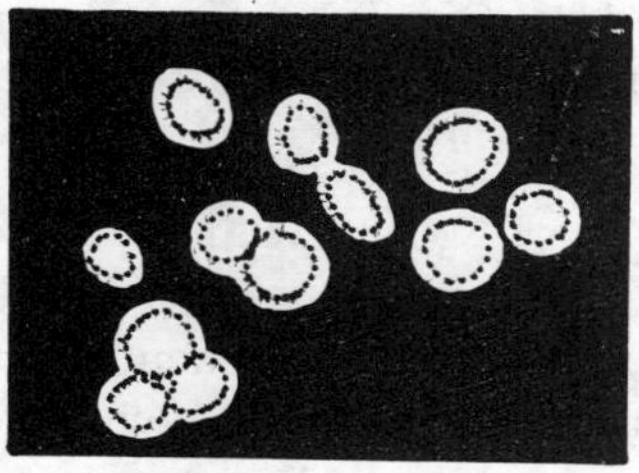

Fig. 5.13 Oxidase reaction on colonies of *N. gonorrhoeae* culture on chocolate agar.

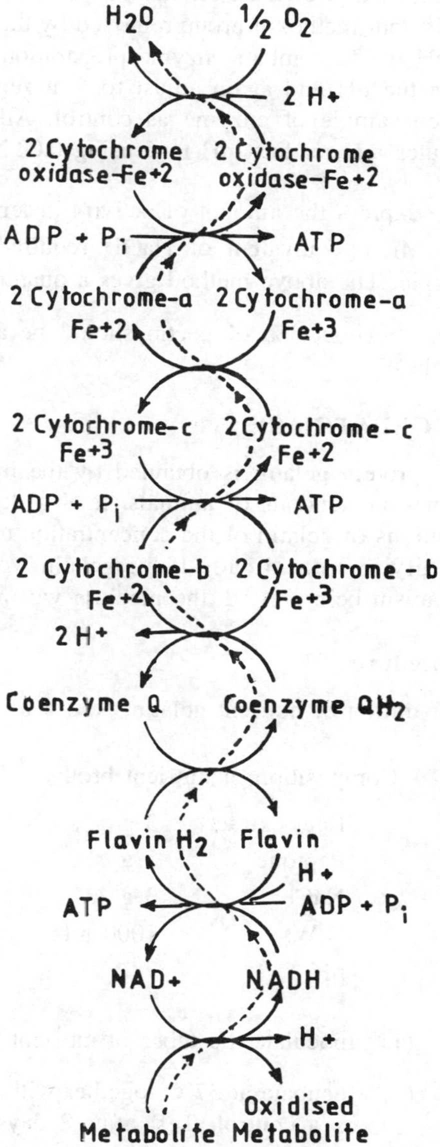

Fig. 5.14 The electron transport system. The dashed line indicates the flow of electrons.

11. PECTINASE

Pectin methyl esterase an enzyme used in hydrolysis of pectin which is expressed in micro equivalent methoxyl group removed by the enzyme sample. To 5 ml of pectin (1%) adjusted to pH 6 add 2 ml of enzyme preparation (culture broth) mix and allow to react at 30°C. Note the pH and again adjust to 6 at zero hour with the help of 0.02 N NaOH. Use the boiled sample of enzyme as control. Allow the enzyme substrate to react for about 10 minutes and readjust pH to 6 using 0.02 N NaOH.

Express the amount of activity in terms of unit ml^{-1}.

Micro equivalent of NaOH required to maintain the original pH hr^{-1} ml^{-1} enzyme sample. The above method gives a quantitative estimation of pectinase.

Note : About 1% of pectin should be added to culture broth so as to induce pectinase synthesis.

12. GELATINASE

The protein gelatin is obtained by the hydrolysis of collagen a component of connective tissue and tendons of animals. It also demonstrates a test for proteolytic enzymes. Water solutions of gelatin of the concentration to be used here are liquid at room temperature and solidify in an acid bath. If the gelatin would have been hydrolysed by the action of micro organism being tested, the medium will remain liquid.

Procedure

Preparation of nutrient gelatin. Add 4% of gelatin in nutrient broth.

Composition of nutrient broth

Beef extract	3 g
Peptone	5 g
NaCl	3 g
D.W.	1000 ml
pH	7.0

(1) Inoculate the tubes of nutrient gelatin with the unknown culture.

(2) Incubate at 37°C together with a sterilised tube of nutrient gelatin that will serve as a control. Test after 2 days and upto 7 days or until a positive reaction is obtained.

(3) To examine hydrolysis, chill the tubes in ice water. The control tubes and tubes in which no hydrolysis has taken place will solidify. Hydrolysed gelatin will remain fluid.

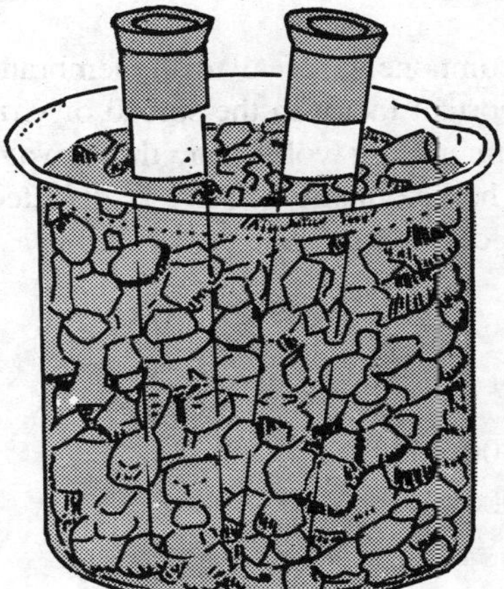

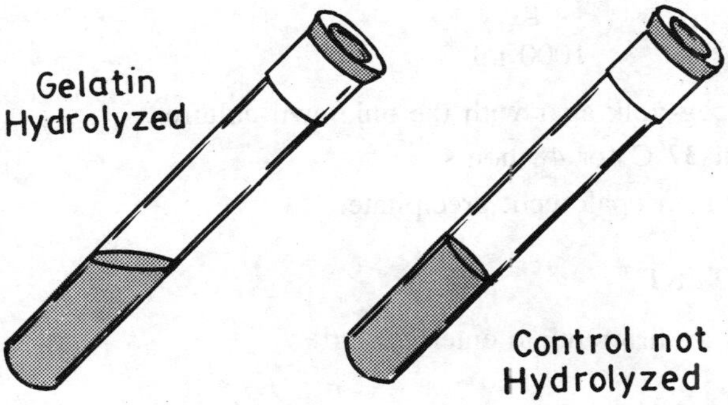

Fig. 5.15 Testing for gelatin hydrolysis.

13. LIPASE

Two common lipids decomposed by microorganisms are triglycerides and phospholipids. The triglycerides are esters of glycerol and fatty acids. Reaction involved in their hydrolysis occur as follows :

$$\begin{array}{c} CH_2\text{–O–C–R} \\ | \\ CH\text{–O–C–R} \\ | \\ CH_2\text{–O–C–R} \\ \text{a triglyceride} \end{array} \xrightarrow[\text{lipase}]{+\,3H_2O} \begin{array}{c} CH_2OH \\ | \\ CHOH \\ | \\ CH_2OH \\ \text{glycerol} \end{array} + \begin{array}{c} RCOOH \\ | \\ RCOOH \\ | \\ RCOOH \\ \text{3 molecules of fatty acids} \end{array}$$

Phospholipids are major components of all cell membranes, the ability to hydrolyse host cell posphòlipids is an important factor in the spread of virulent microorganisms. It can be tested by growing the culture in egg yolk agar, the clevage of phosphate ester bonds form a water insoluble lipid. The enzymatic activity is detected by a zone of opalescence in the medium surrounding the cell mass.

Procedure

Egg Yolk Agar

Egg yolk	100 ml (5%)
Plate count agar	900 ml

Plate Count Agar

Tryptone	5 g
Yeast extract	2.5 g
Glucose	1 g
Agar	15 g
D.W.	1000 ml

(1) Inoculate egg-yolk agar with the unknown culture.

(2) Incubate at 37°C for 48 hours.

(3) Observe for an opalescent precipitate.

14. GLUCONATE TEST

The test is employed to differentiate enterobacteria.

Gluconate Broth

Yeast extract	1.0 g
Peptone	1.5 g
K_2HPO_4	1.0 g
Potassium gluconate	40.0 g
or Sodium gluconate	37.25 g
D.W.	1000 ml

Dissolve by heating, adjust pH to 7.

Filter, if necessary.

Composition of Benedict's reagent

Copper sulfate	18 g
Sodium carbonate	100 g
Sodium citrate	200 g

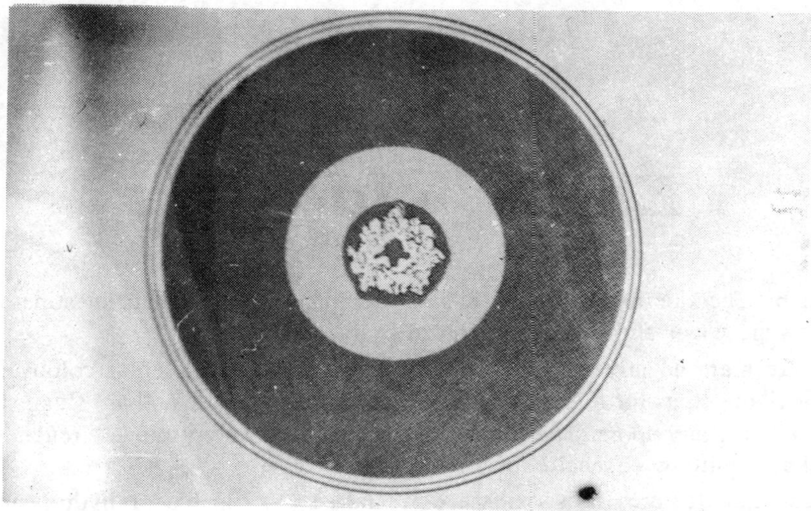

Fig. 5.16 Lipid hydrolysis.

KCNS	125 g
Potassium ferrocyanide solution	(5%) 5.0 ml
D.W.	100 ml

(1) Inoculate gluconate broth.

(2) Incubate for 48 h at 37°C.

(3) Add an equal volume of Benedict's reagent and place in a boiling water bath for 10 mins.

(4) An orange or brown precipitate indicates gluconate oxidation. A deep blue colour is negative response.

6

Escherichia coli

Discovered by Theodor Escherich in 1885 as a commensal of human intestine and is also found in sewage water and soil contaminated by faecal matter.

These are gram negative rods 2 to 4 μ by 0.4 μ commonly seen as colobacillary form rarely as air filamentous form. These have 4-8 peritrichate flagella and are sluggishly motile. They do not form any spore. The facultative anaerobes develop into a circular raised and smooth colony emitting a faecal odour.

Besides these features it is oxidase (–) catalase (+) capable of hydrolysing gelatin, reduces nitrate and can produce indole. It is incapable of utilising citrate and urea.

ISOLATION

Material and Methods

The genus *Shigella* and *Escherichia* are closely related except for the inability of former to utilise Lactose – the **Bromothymol blue** agar is used as a selective medium.

Medium Composition

Peptone	10 g
NaCl	5 g
Oxoid yeast extract	5 g
D.W.	1000 ml
Agar	20 g
pH	7

Autoclave it and add following sterilised solutions.

5% Maranil solution	1 ml
50% Sodium thiosulfate	2 ml
1% Bromothymol blue	10 ml
33% Lactose	25 ml
33% Glucose	1.2 ml

(i) Prepare serial dilution* of 5 ml of the collected sample

(ii) Inoculate 0.1 ml of highest dilution on the agar plate.

* Serial dilution : Method discussed in Chapter 3. Cultivation of Bacteria.

If the above method is difficult to perform. Use standard analysis of water.

STANDARD ANALYSIS OF WATER

Fig. 6.1 The presumptive test for the coliform group showing gas production from lactose broth.

It consists of 3 steps :

(i) Presumptive test

(ii) Confirmed test

(iii) Completed test

Presumptive Test

1. Inoculate 10 ml portions of each water sample collected to 10 ml of lactose broth (double strength)
2. Incubate at 37°C for 48 hrs.
3. Observe after 24 hrs, presence of gas shows a positive presumptive test. Appearance of bubble in next 24 hrs shows a doubtful presumptive test. The absence of gas formation indicates a negative confirmed test.

   Composition of Lactose Broth

   Beef extract 6 g

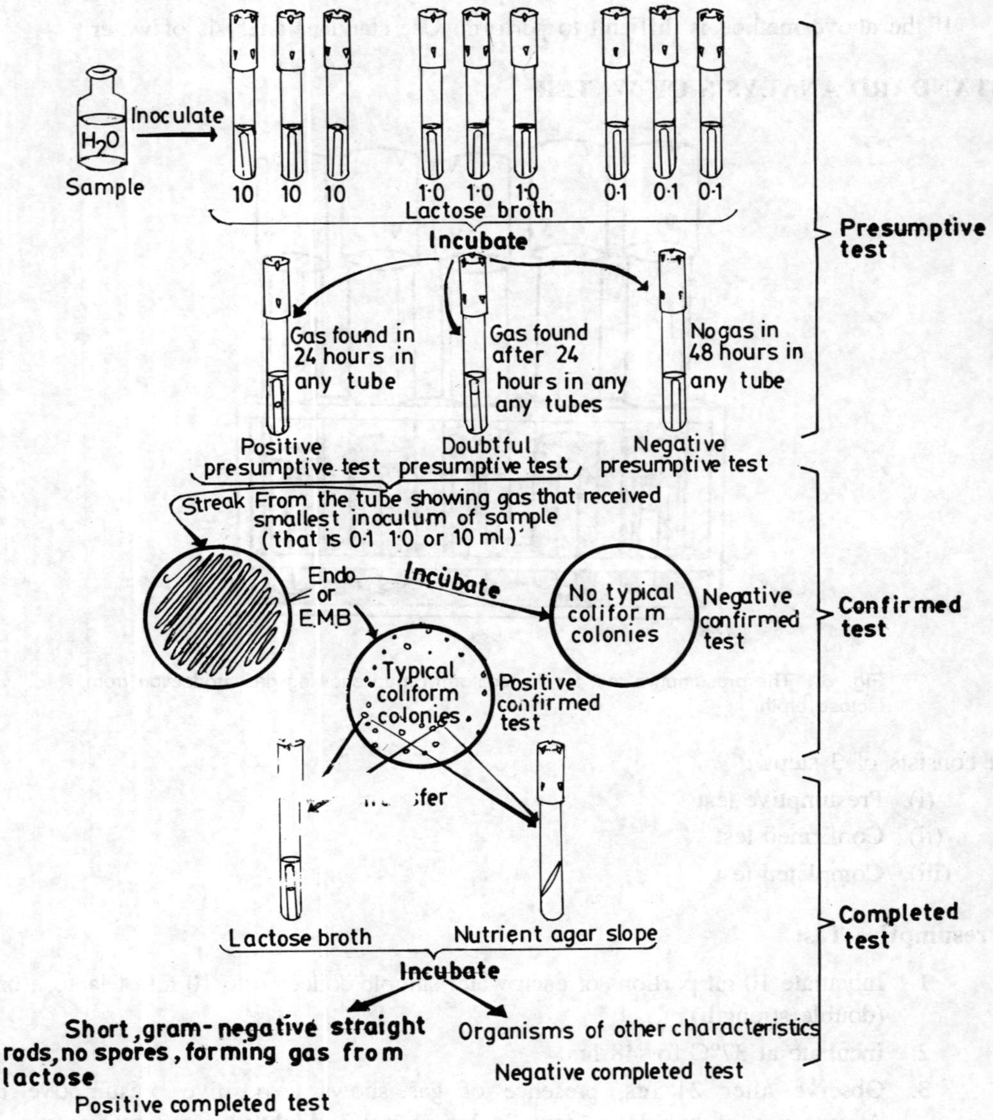

Fig. 6.2 Flow diagram for the standard analysis of water.

Peptone	10 g
Lactose	10 g
D.W.	1000 ml

Confirmed Test

1. Streak, using a very little inoculum, a plate of Eosine Methylene Blue agar, or Endo agar.
2. Inoculate about one loopful of culture from positive tube to Brilliant Green Lactose Broth.
3. Incubate at 37°C for 48 hrs.

Gas formation in lactose broth shows a positive confirmed test. While appearance of characteristic small, dark greenish or large pinkish mucoid colonies also confirms the presence of coliforms in the sample.

Eosine Methylene Blue Agar (EMB)

Peptone	10 g
Lactose	5 g
Sucrose	5 g
K_2HPO_4	2 g
Eosin Y	0.4 g
Methylene blue	0.06 g
Agar	15.0 g
D.W.	1000 ml

Endo Agar

Peptone	10 g
Lactose	10 g
K_2HPO_4	3.5 g
Na_2SO_3	5 g
Basic fuchsin	0.5 g
Agar	15 g
D.W.	1000 ml

Brilliant Green Lactose Bile Broth (BGLB)

Composition	10 g
Lactose	10 g
Brilliant green	13.3 mg
Distilled water	1000 ml
Final pH	7.2

Completed Test

1. Inoculate a loopful of bacterial cultural from EMB or ENDO AGAR PLATE to Lactose broth.
2. Streak a loopful of culture on to nutrient agar slant.
3. Incubate at 37°C for 48 hrs.
4. Gas formation in Lactose broth and the appearance of gram negative bacteria on the plates shows a positive completed test.

Identification

The isolate tested Gram negative and picked from nutrient agar slant should be subjected to following tests.

(A) Morphological Tests*

1. Flagella stain
2. Capsular stain
3. Gram stain
4. Endospore stain
5. Colony Morphology
6. Motility

(B) Biochemical Tests*

1. Catalase
2. Oxidase
3. Gas production from glucose
4. Acid production from mannitol and sucrose
5. Gelatin Hydrolysis
6. H_2S from T.S.I.
7. Urease
8. IMViC Reaction
9. Nitrate reduction

* Methods of all of the above tests are given in Chapter 4 and 5.

Discussion

A smooth margined, circular colony with elevated surface, each member showing peritrichous flagella, but devoid of spores and emitting a faecal odour could be taken as a member of Enterobacteriaceae. This can be confirmed by acid production in glucose broth, indole production and nitrate reduction.

A deep red colour on addition of methyl red confirms the coliform nature, the positive catalase test and a negative oxidase test shows that there is poor cytochrome oxidase activity.

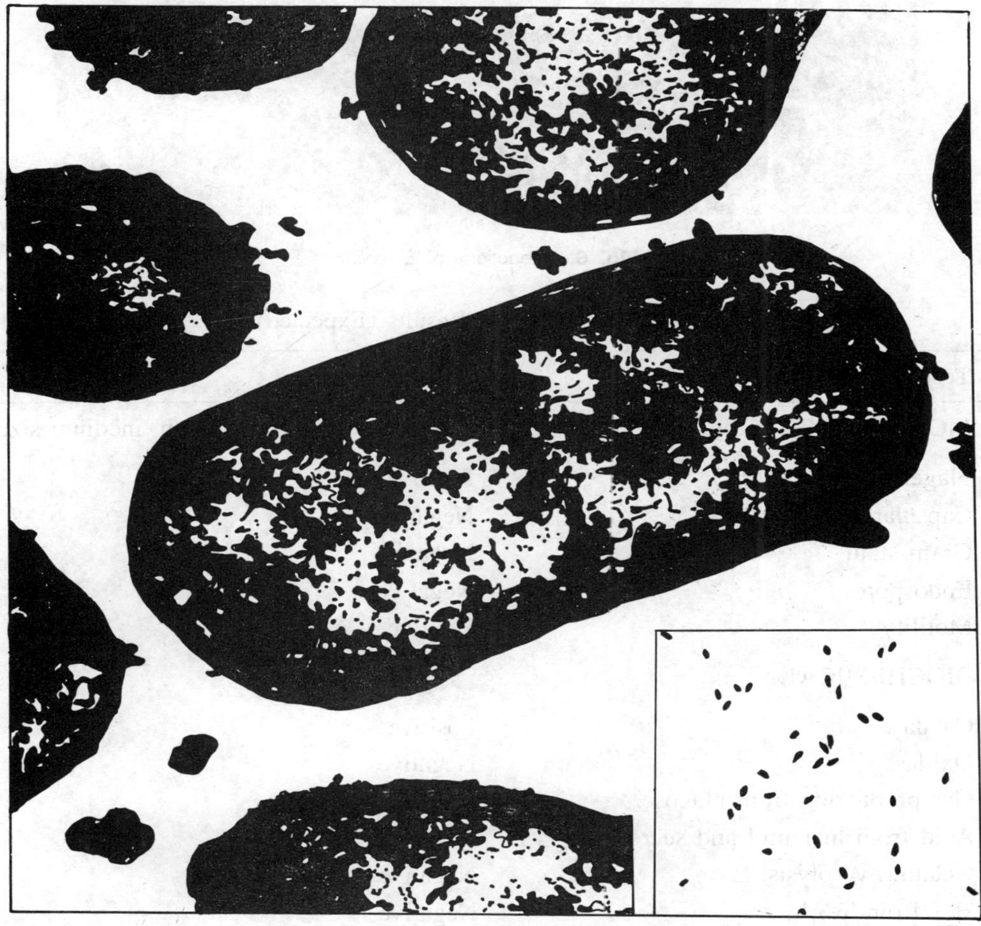

Fig. 6.3 Bacterium *Escherichia coli.*

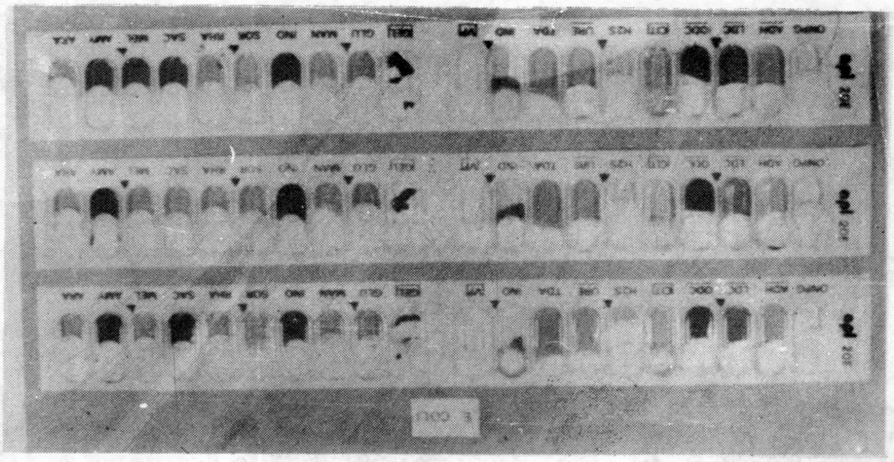

Fig. 6.4 Reactions of *E. coli*.

Table 7 Summary of Results (Expected)

	Test	Result
	MORPHOLOGICAL COLONY	Circular, raised and smooth, medium sized
1.	Flagella stain	Positive
2.	Capsular stain	Negative
3.	Gram stain	Negative
4.	Endospore	Negative
5.	Motility	Positive
	BIOCHEMICAL	
1.	Catalase	Positive
2.	Oxidase	Negative
3.	Gas production from glucose	Positive
4.	Acid from mannitol and sucrose	Positive
5.	Gelatin hydrolysis	Negative
6.	H_2S from T.S.I.	Negative
7.	Urease	Negative
8.	IMViC reactions	
	(a) Indole production	Positive
	(b) Methyl Red Test	Positive
	(c) Voges-Proskauer Test	Negative
	(d) Citrate Utilisation	Negative
9.	Nitrate Reduction	Positive

Results

The isolate picked from bromothymol blue agar or from the nutrient agar slant, last step of standard analysis of water should show a circular, raised and smooth colony, faecal odour. A negative response to Gram reaction, Oxidase test, H_2S formation from triple sugar iron agar, Indole production and positive methyl red reaction.

Appearance of pink bacteria in a gram reaction indicates low content of peptidoglycan in cell wall.

A bipositive and binegative response to IMViC reactions accompanied with above morphological characteristics leaves no doubt about the isolate being *Escherichia coli*.

7

Pseudomonas

Pseudomonas is a very large and important group of non fermenting, gram negative bacteria which is found in substantial numbers as free living saprophytes in soils, fresh water, and many other natural habitats. Some species are pathogenic to humans, animals or plants. The basic morphological characters common to *Pseudomonas* are the rod shaped cells; usually motile, the presence of flagella, a gram negative reaction and the absence of spores. *Pseudomonas* is one of the few microorganisms which can grow in simple mineral media at the expense of large varieties of low molecular weight carbon compounds. Identification is based on the common morphological and biochemical reactions, motility, the types of flagella, oxidase test, catalase test, presence of extracellular amylase, gelatinase and caseinase, oxidation fermentation test, nitrate reduction and production of pigments.

Pseudomonas isolation using Enrichment technique

(i) Pre-incubate the source material at 20°C for 24 hrs. Solid sources, e.g., eggs, fish or poultry should be placed in phosphate buffer dilution tanks, shaken and incubated at 20°C for 24 hrs. This facilitates their distribution during streak plating.

(ii) Following pre-incubation, streak the source material on Novobiocin-penicillin cycloheximide agar.

Composition of Novobiocin-penicillin cycloheximide agar g/l.

Novobiocin	45 μg
Penicillin G	44.9 mg (75,000 units)
Cycloheximide	75 mg

Mix these in 3 ml ethanol dilute with 50 ml of sterilised distilled water. Add to 1 litre sterilised, tempered *Pseudomonas* agar F.

Pseudomonas produce pyocyanin (water soluble) and fluorescein pigment. The latter fluoresce when exposed to uv light. These pigments are produced on P agar and F agar medium.

Materials and Methods

I. Isolation

(i) Direct method

Nutrient agar medium

NaCl	3 g
Beef extract	3 g
Peptone	5 g
Agar	15 g
D.W.	1000 ml

(i) Inoculate 1 g of compost soil in about 10 ml of sterilised distilled water.

(ii) Prepare a serial dilution as discussed in Chapter 3 (cultivation of bacteria).

(iii) Spread 0.1 of highest dilution on nutrient agar plate. It is advised that the sample must first be inoculated on *Pseudomonas* P agar, *Pseudomonas* F agar and *Pseudomonas* cetrimide agar and then transferred to nutrient agar so that the contamination can be checked.

Identification

Perform following Morphological and Biochemical Tests.

MORPHOLOGICAL TESTS

1. Colony Morphology
2. Flagella Stain
3. Gram Stain
4. Endospore Stain
5. Pigment Production
6. Motility

BIOCHEMICAL TESTS

1. Catalase
2. Oxidase
3. Gas Production in glucose Broth
4. Gelatinase
5. Caseinase
6. Oxidation/Fermentation Reaction
7. Amylase
8. H_2S Production
9. Nitrate reduction
10. Growth at 42°C.

DICHOTOMOUS CHART FOR IDENTIFICATION OF *Pseudomonas* SP.

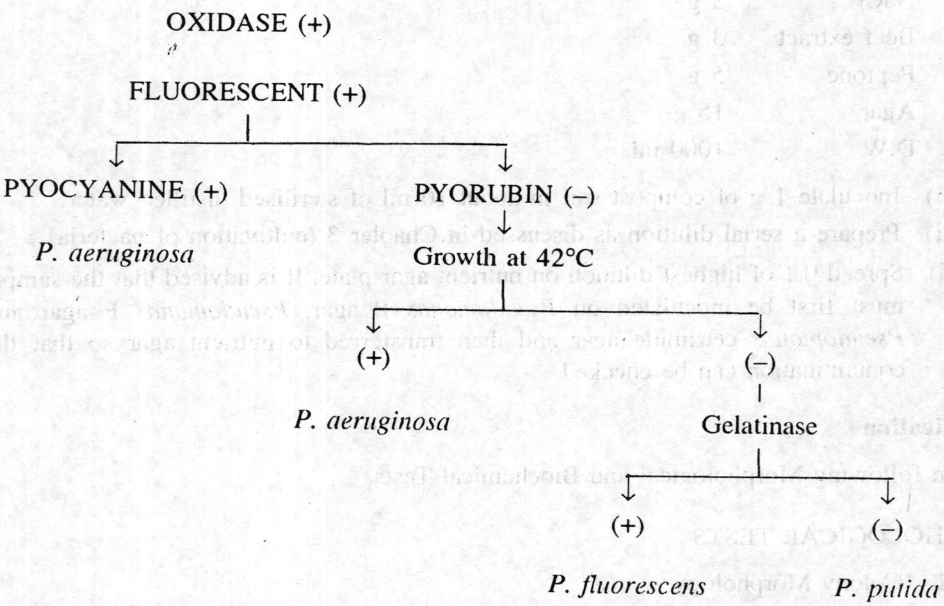

Pseudomonas P agar (pigment production)

Peptone	20 g
Potassium sulfate	10 g
$MgCl_2$	1.4 g
Agar	15 g
D.W.	1000 ml
pH	7

Pseudomonas F agar

Peptone	10 g
Proteose peptone	10 g
Dipotassium phosphate	1.5 g
Magnesium sulfate	1.5 g
Agar	15 g
D.W.	1000 ml
pH	7

Method of Preparation

1. Suspend above ingredients in a litre of distilled water and add 10 g of glycerol.
2. Heat the medium to boiling in order to dissolve completely.
3. Dispense in tubes or flasks and autoclave.
4. Allow to cool in slanted position.
5. Dispense sterilised medium onto sterilised pertriplates.
6. Inoculate tubes and plates both by surface streaking.

Composition of cetrimide agar (for isolation of *P. aeruginosa*)

Peptone	20 g
$MgCl_2$	1.4 g
Potassium sulfate	10 g
Cetrimide (Cetytrimethyl diammonium Bromide)	0.3 g
Agar	13.6 g
D.W.	1000 ml

Method

1. Suspend above ingredients in 1 litre distilled or de-ionised water. Add 10 ml glycerol and heat to boiling to dissolve.
2. Dispense in tubes or flasks.
3. Sterilise in the autoclave for 15 minutes at 15 lbs pressure.

Results

The round shaped (small dot) smooth margined colony isolated from nutrient agar plate should be gram negative, motile with rod shaped cells, devoid of spore, sheath or flagella. Positive to catalase and oxidase reactions.

The colony morphology is smooth margin, round pale brown or cream colour negative reaction with glucose broth. Negative gelatinase, caesinase and H_2S production is expected.

Discussion

The genus *Pseudomonas* is characterised by morphological properties such as the rod shaped cells which are either straight or slightly curved in one plane, the presence of polar flagella, a gram negative reaction, the absence of spores, sheaths and prosthecas, and by physiological properties such as a nutrition of the chemoorganotrophic metabolism typical respiratory, an absence of fermentation and photosynthesis, an inability to fix dinitrogen, and a capacity for growth at the expense of a large variety of organic substances. Catalase reaction (positive) and glucose broth with bromo thymol blue indicator and durham tube (no change in the dye-colour by acid fermentation products and gas production) indicate that the isolate has an aerobic, obligately respiratory type of metabolism with oxygen as the terminal electron

Table 8 Summary of Results (Expected) Morphological and Biochemical Reactions

S. No.	Tests	Results
MORPHOLOGICAL		
1.	Colony Morphology	Smooth, Round
2.	Flagella stain	Positive
3.	Gram stain	Negative
4.	Endospore stain	Negative
5.	Motility	Positive
6.	Pigment Production	Positive
BIOCHEMICAL		
1.	Catalase	Positive
2.	Oxidase	Positive
3.	Gas Production in glucose Broth	Negative
4.	Gelatinase	Negative
5.	Caseinase	Negative
6.	Oxidation/Fermentation Reactions	Oxidation
7.	Amylase	Negative
8.	H_2S Production in T.S.I.	Negative
9.	Nitrate Reduction	Negative
10.	Growth at 42°C	Positive

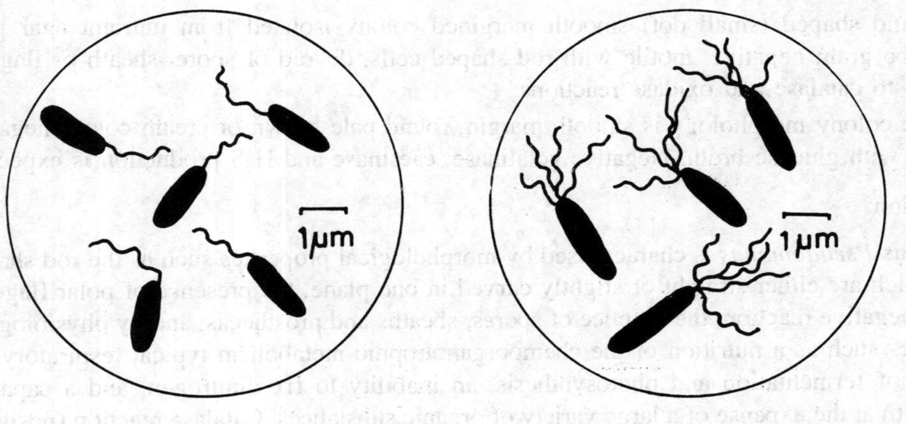

A Pseudomonas aeruginosa B. Pseudomonas fluorescens

Fig. 7.1.

acceptor. The presence of cytochrome C in the electron transport chain is ascertained by positive oxidase reaction. "O" reaction in the oxidase/fermentation test indicates that the isolate catabolised the glucose by oxidation. The production of fluorescent pigment is a definite reaction for *Pseudomonas*. The function of these pigments is to chelate iron in the environment for the use of the cells, and its overproduction is stimulated when the culture is stressed to iron poor medium, of 100 sp of *Pseudomonas* only 8 produce fluorescent pigment. These 8 sp. are : *P. aeruginosa, P. aureofacians, P. chololoraphis, P. fluorescens, P. putida.* Among plant pathogens – *P. syringae, P. viridiflava, P. chichorii.*

According to Bergey's Manual *P. putida* has synonyms such as *P. convexa, P. eisenbergii, P. Incognita, P. ovalia, P. rugosa* **and** *P. striata.* It was isolated from soil and water. Its G + C content is 62.5 mole %.

8

Rhizobium

Rhizobia are short to medium, gram negative rods (0.5-0.9 × 1.2-3 μm) older cells bear capsule, and are sporulated, flagella are arranged in polar, subpolar or peritrichous fashion.

The growth of most *Rhizobia* is water clear, white or cream coloured, with transparent or translucent areas differentiated into regions of greater opacity owing to the presence of a strain or substrain with in a species. Identification of *Rhizobium* is based on common biochemical tests such as : Gram reaction, motility, flagellar staining, no H_2S production, litmus milk reaction, inability to utilise citrate, and intolerance to 2% NaCl on YMA.

Materials and Methods

ISOLATION

1. Composition of Yeast Extract Mannitol Agar (YMA)

K_2HPO_4	0.5 g
$MgSO_4 . 7H_2O$	0.2 g
NaCl	0.1 g
Mannitol	10 g
Yeast Extract Powder	0.4 g
Agar	15 g
D.W.	1000 ml

Identification

S. No.	Tests
A	MORPHOLOGICAL
1.	Colony
2.	Gram Stain
3.	Flagella Stain
4.	Endospore Stain

5. Capsular Stain
6. Motility

B BIOCHEMICAL

1. Growth on Glucose Peptone Agar
2. Litmus-Milk Test
3. Growth on 2% NaCl + YMA
4. H_2S Production
5. Citrate Utilisation
6. Nodule formation
7. Growth at pH 9.5
8. Ability to use Dulcitol and Raffinose
9. Acid production
10. Biuret utilisation. 2% Biuret in YMA
11. Response to thiamine and pantothenate 0.01 g in Yeast Extract Mannitol Broth

---

Composition of Glucose-Peptone Agar

Glucose	10 g
Peptone	20 g
NaCl	5 g
Agar	15 g
pH	7.2
D.W.	1000 ml

Direct Isolation

(1) Take 1 g of soil (collected from legume inhabiting field) and dissolve it in about 10 ml of D.W.

(2) Prepare serial dilutions as discussed in Chapter 3 (cultivation of bacteria).

(3) Inoculate about 0.1 ml of the highest dilution on YMA plates.

(4) Examine the plates daily for 5 days at 26°C for the development of typical white colonies either :

(i) with fast grower 2.5 mm viscous or with free flowing gum (exo-polysaccharide);

(ii) a slow grower detectable after five days and not much greater than 1-1.5 mm. On 10 days incubation these will be gum resistant and adherently viscous.

Note : Inclusion of congo red dye may help recognition of relatively uncoloured rhizobial colonies avoiding the chance of mistaking *Agrobacterium* (deep pink/red) for *Rhizobia* which gives all the biochemical tests designed for *Rhizobia.*

Addition of 0.002% of cycloheximide eliminates the fungal contaminants, if any.

NODULE FORMATION

The principle is to grow sterile host plants in sterilised nitrogen free medium with and without the *Rhizobium* strain under test and to observe whether nodules are formed in presence of bacteria. The test can also show whether or not nitrogen is fixed in nodules.

Small seeded plants (e.g. *Trifolium pratense*) can be grown wholly enclosed with in glass tubes fitted with cotton wool plugs to protect from contamination.

Clear uniform seed is rinsed with 95% ethanol and immersed for 3 min in 0.2% (w/v) $HgCl_2$ solution. The seed is then washed thoroughly with at least 10 changes of sterile water and planted directly on agar slopes in tubes or set to germinate on moist filter paper or water agar (1.2%) in petridishes at an appropriate temperature.

Composition of seedling agar

CaH_2PO_4	1 g
K_2HPO_4	0.2 g
$MgSO_4 . 7H_2O$	0.2 g
NaCl	0.2 g
$FeCl_3$	0.1 g
Agar	8-15 g
D.W.	1000 ml
pH	6.8

Inoculate the slopes with bacteria.

Nodules develop when the plants are 2-4 weeks old and first tree leaves open.

Results

Large white colonies, isolated from soil on yeast extract mannitol agar with gram negative character, ability to grow at pH 9.5 and utilise biuret. A positive flagella staining reaction and nodule formation.

Table 9 Summary of Results Expected

Tests	Results
A. MORPHOLOGICAL	
1. Colony	Large, Smooth
2. Gram Stain	Negative
3. Flagella Stain	Positive
4. Endospore stain	Positive
5. Capsular stain	Positive
6. Motility	Positive
B. BIOCHEMICAL	
1. Growth on Glucose Peptone Agar	Slight
2. Litmus milk reaction	Negative
3. Growth on 2% NaCl + YMA	Negative
4. H_2S production	Negative
5. Citrate utilisation	Negative
6. Nodule formation	Positive
7. Growth at pH 9.5	Negative
8. Ability to use Dulcitol and Raffinose	+/–
9. Acid production	+/–
10. Biuret utilisation	+/–
11. Response to thiamine and pantothenate	+/–

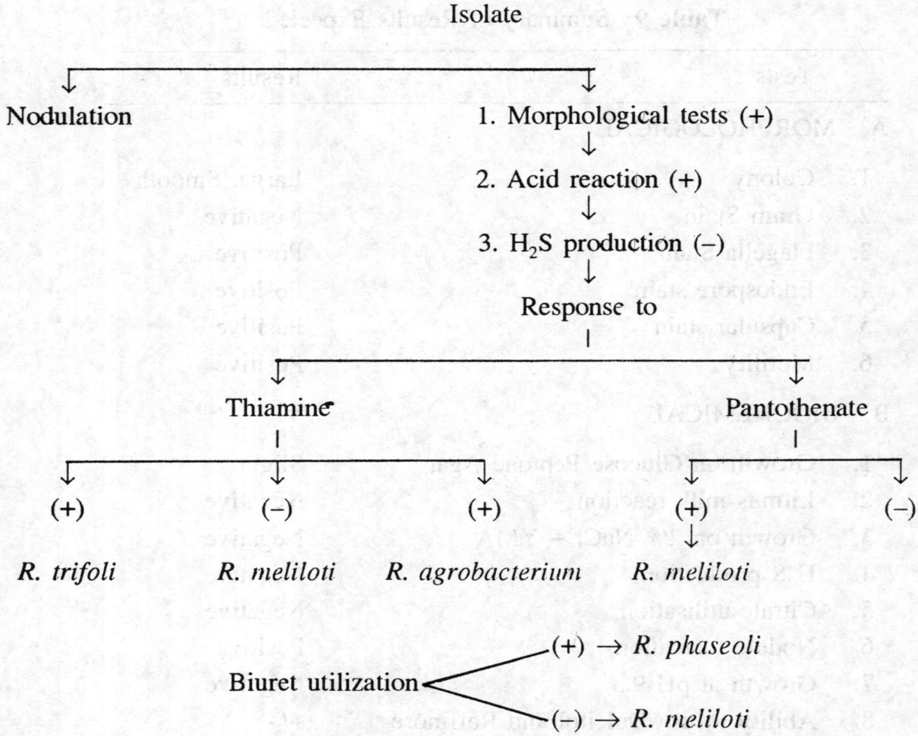

Discussion

The genus *Rhizobium* is characterised by morphological properties such as the rod shaped cells presence of flagella in either polar, subpolar or peritrichous fashion. A gram negative reaction, absence of spores in young culture, however, old cultures show spores. Chemo-organotrophic mode of nutrition, ATP generation through glucose break down, aerobic nature, ability to fix dinitrogen are characters of *Rhizobium*.

The principle species of *Rhizobium* are :

R. trifoli, R. phaseoli, R. meliloti. The key characters of these species are shown in dichotomous chart.

G + C content of *Rhizobium* is 59.1-65.5 mole %.

9

Shigella

Isolated by Kiyoshi Shiga (1898) from a case ot acute bacillary in Japan. *Shigella* is a large group of negative, non-motile, slender, rod shaped organism about (0.5 μ × 2.5 μ) fimbriae present pathogenic to humans and animals and cause Bacillary dysentry.

Physiologically, they are unable to fix nitrogen, fermentative, nonphotosynthetic aerobic or facultatively anaerobic in nature. Isolation is achieved through enrichment method while identification is based on general biochemical tests such as Gram reaction, motility, oxidase, catalase reaction, acid and gas production on glucose, lactose, sucrose and salicin, hydrolysis of gelatin and urea, utilisation of citrate, and other IMVIC reactions etc.

Identification of species involves the fermentation reactions using lactose and mannitol as carbon source.

Materials and Methods

Composition of Mac-Conkey Agar

Dissolve 20 g of peptone, 5 g of sodium chloride, 5 g of sodium tauro cholate in a litre of D.W., assisting the solution by heat. Take pH to 8 with 40% NaOH. Add 20 g of agar and dissolve it by steaming. Clarify with white of egg at 55°C (white of 2 eggs is dissolved in 50 ml of water and added to one litre of the medium and steamed for 2 hrs) filter adjust pH to 7.4. Distribute in 100 ml quantities in flasks. Before use, melt agar and add 10% lactose solution in water, previously sterilised, add 1% final concentration of neutral red (dissolved in 50% alcohol) and pour into sterilised petri dishes under aseptic condition.

Selenite - F broth

Sodium hydrogen (acid) selenite	4 g
Peptone	5 g
Lactose	5 g
Na_2HPO_4	9.5 g
NaH_2PO_4	0.5 g
pH	7.2
D.W.	1000 ml

Distribute in 10 ml quantities in test tubes and sterlise by steam for 30 mins. A slight amount of precipitate may occur but it will not affect the medium.

Direct Isolation

1. Inoculate about 1 g of sewage sample in 10 ml of distilled water.
2. Prepare a serial dilution of the suspension.
3. Spread 0.1 ml of highest dilution on the Mac-Conkey/EMB agar plate.
4. Incubate for 2 days at 37°C.
5. Restreak on to nutrient agar plate to check contamination.
6. Repeat the procedure till a single type of colony is obtained.

Enrichment Isolation

1. Repeat procedure (1) to (2) of Direct Isolation.
2. Inoculate 0.1 ml of above dilution in selenite - F broth
3. Incubate for 2 days at 37°C.
4. Take a loopful and streak on Mac-Conkey agar plate.
5. Repeat procedure (4) till a single of colony is obtained.

Fig. 9.1 Division of *Shigella*.

Subject the isolate to identification.

Identification

Tests

A. MORPHOLOGICAL

1. Colony
2. Gram Stain
3. Flagella Stain
4. Endospore Stain
5. Capsular Stain

B. BIOCHEMICAL

1. Catalase
2. Oxidase
3. Gas from glucose broth

4. Gas from lactose Broth
5. Gas from Mannitol
6. Gelatin Hydrolysis
7. H_2S from T.S.I.
8. Indole production
9. Simmon's citrate utilisation
10. Acid production using salicin
11. Gluconate test
12. V.P. Test
13. Nitrate Reduction

IDENTIFICATION OF SPECIES

DICHOTOMOUS CHART FOR IDENTIFICATION OF SPECIES

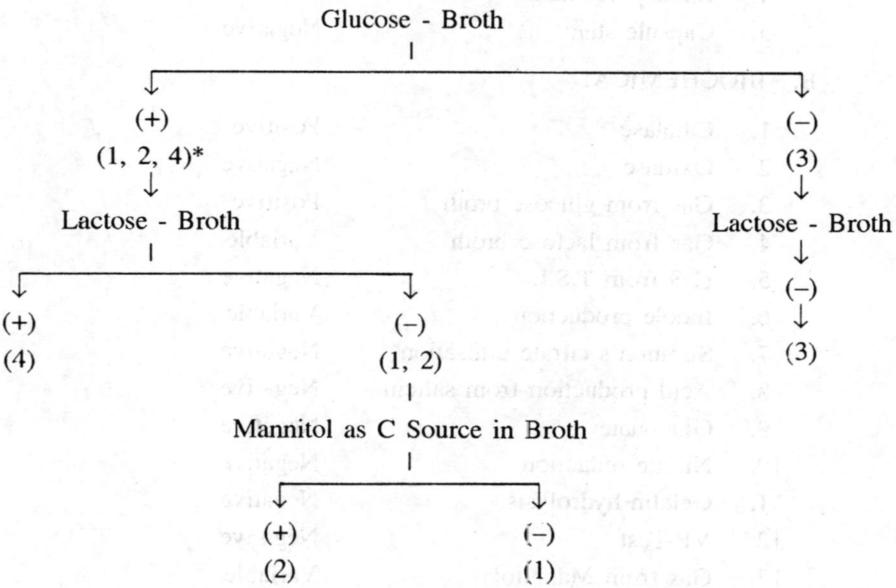

* 1. *Shigella bacilli*
2. *Shigella flexneri*
3. *Shigella boydii*
4. *Shigella sonei*

Results

A translucent colony picked up from Mac Conkey agar (a selective media for *Shigella*) bearing rod shaped cells without spores and flagella stand as a prominent candidate for the above mentioned genus. The positive response to catalase and oxidase test, inability to form acetyl methyl carbinol (V-P negative), utilise citrate, unable to ferment salicin are some of the important characters of above genus.

Table 10 Summary Results (Expected)

S. No.	Test	Results
A.	MORPHOLOGICAL	
1.	Colony	Large, Translucent
2.	Gram stain	Negative
3.	Flagella stain	Negative
4.	Endospore stain	Negative
5.	Capsule stain	Negative
B.	BIOCHEMICAL	
1.	Catalase	Positive
2.	Oxidase	Negative
3.	Gas from glucose broth	Positive
4.	Gas from lactose broth	Variable
5.	H_2S from T.S.I.	Negative
6.	Indole production	Variable
7.	Simmon's citrate utilisation	Negative
8.	Acid production from salicin	Negative
9.	Gluconate	Negative
10.	Nitrate reduction	Negative
11.	Gelatin hydrolysis	Negative
12.	VP-Test	Negative
13.	Gas from Mannitol	Variable

Discussion

A metabolism typical respiratory, absence of photosynthesis. Ability to ferment sucrose, glucose and mannose, absence of spores and flagella, inability to fix dinitrogen, chemo-organotrophic mode of nutrition are characteristic of these unsheathed bacilli.

Growth on E.M.B. agar but inability to utilise simmon's citrate and ferment salicin confirms the *Shigella* nature of the organism.

Negative response to nitrate reduction shows that the isolate is unable to use nitrate (NO_3^-) as terminal electron acceptor instead of using oxygen, while the aerobic nature shows the obligately respiratory type of metabolism. The above characters place the isolate in the genus *Shigella.*

The identification of species is based on the ability to ferment various carbon sources since *Shigella bacilli* can ferment only glucose while *Shigella sonei* can ferment all the three glucose, lactose and mannose. *Shigella boydii* is very distinct in having tendency to ferment indole and glucose broth while *Shigella flexneri* is limited to glucose and mannitol only (see table).

Table 11 Key Characters of *Shigella* sp.

Carbon	*Shigella flexneri*	*Shigella boydii*	*Shigella sonei*	*Shigella bacilli*
Glucose	+	–	+	+
Mannitol	+	+	+	–
Lactose	–	–	+	–
Indole	–	+	–	–

According to Bergey's manual *Shigella* has been isolated from faeces. The G + C content ranges from 49-53 mole %. Among the 4 species of *Shigella* each has its own serotype.

10

Proteus

The genus *Proteus* belongs to the tribe Proteae, a member of Enterobacteriaceae. They occur frequently in pathological specimens submitted for bacteriological examinations. The ability to swim over the surface of agar media, led to its name *Proteus* (a Greek deity, *Proteus* which could change its shape).

Proteus is characterised by motile, gram negative bacilli, highly pleomorphic with spreading colonies (spreaders) actively proteolytic, but unable to ferment lactose, ferment other carbohydrates with acid and gas production.

These are found in sweage, meat and manure and have the dimension of $0.5\ \mu \times 0.2\ \mu$ as short rods while upto $20\ \mu$ as filamentous forms. Spore and capsule absent, fimbriated with a facultative aerobic nature.

Materials and Methods

I. Isolation

1. Take 5 ml of sewage sample.
2. Prepare serial dilutions as discussed in Chapter 3.
3. Spread 0.1 ml each of the above dilutions and plate on to the previously made nutrient agar plates.
4. Transfer any swarming colony if visible to Deoxycholate citrate agar.
5. Observe the plate for isolated colonies after incubation of 48 hrs at 37°C. Swarming should stop by now.
6. Check the contamination by restreaking on nutrient agar plates. The colonies should now gain their swarming nature.

Deoxycholate citrate agar is used for *Proteus* enrichment.

Composition of Deoxycholate citrate agar

Peptone	10 g
Lactose	10 g
Sodium deoxycholate	1 g
NaCl	5 g
K_2HPO_4	2 g

Ferric citrate	1 g
Sodium citrate	1 g
Neutral red	0.03 g
Agar	15 g
D.W.	1000 ml

PPA Slope Medium (Ewing et al., 1957)

Dissolve 2 g of DL-phenylalanine; 3 g meat extract 1 g Na_2HPO_4, 5 g NaCl and 1.2 g agar in 100 ml of D.W.

Tube sterilise at 10 lbs for 10 mins and solidify in a long slant. Inoculate the whole slant heavily and incubate overnight. Add 4-5 drops of 10% (w/v) ferric chloride to the slope and allow to run over the surface. Observe the production of dark green colour.

The isolate is subjected to following tests for identification.

S. No. Tests

A. MORPHOLOGICAL

1. Colony
2. Gram stain
3. Flagella stain
4. Capsule stain
5. Endospore stain
6. Motility

B. BIOCHEMICAL

1. Urease
2. Indole production
3. H_2S in T.S.I.
4. Citrate Utilisation
5. Gas production in :
   a. Glucose, Lactose
   b. Mannitol
   c. Salicin
6. PPA slope Test
7. Voges - Pros Kauer Test
8. PPA Liquid Test
9. Gelatinase

PPA Slope Test (Phenyl-Pyruvic Acid Test) Henrickson (1950)

Phenylalanine Solution

Dissolve 0.2 g DL-phenylalanine in 100 ml D.W. Add 0.1 g merthiolate and 4 ml of 0.01% of phenol red. Turn it to pink colour (if necessary by adding 0.1 M sodium carbonate).

Method : Suspend the growth of an overnight agar culture in normal saline, transfer to a tube add 0.5 ml of phenylalanine solution. Shake the mixture well and incubate the tube in horizontal position for 4 hrs, add 10% H_2SO_4 till the colour changes from pink to yellow (pH 2). Add about 6 drops of half saturated ferric ammonium sulfate Fe (NH_4) $(SO_4)_2$. Shake thoroughly and observe for the development of a deep green colour, when the test is positive. The depth of the colour indicates the degree of the positivity of the reaction.

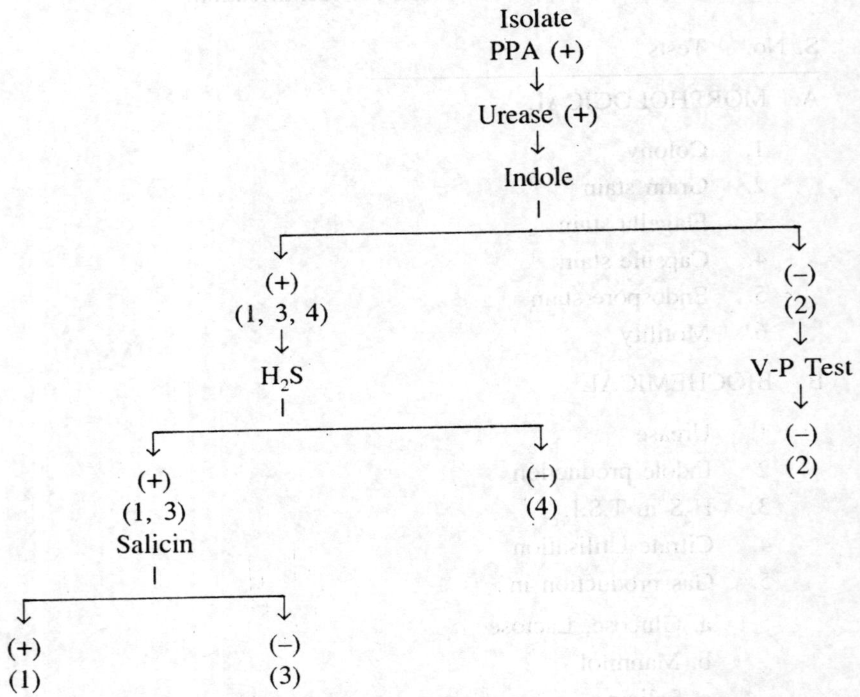

1. *Pr. vulgaris*
2. *Pr. mirabilis*
3. *Pr. morganii*
4. *Pr. rettgeri*

Results

Irregularly shaped, swarming colony which turns to small isolated ones if isolated from sewage or manure can be taken as bacterium *Proteus*. These are small, gram negative, motile with small rod like structure. No spore sheath or capsule is seen. Each cell is however fimbriated, flagellar staining is positive.

Genus is a non-lactose fermenter but can produce acid and gas from glucose, maltose, salicin, mannitol etc. depending upon species.

Spreading on nutrient agar stops in presence of antiseptics, 5% alcohol.

Table 12 Summary of Results (Expected)

S. No.	Tests	Results
MORPHOLOGICAL		
1.	Colony	Pleomorphic
2.	Gram stain	Negative
3.	Flagella stain	Negative
4.	Capsule stain	Negative
5.	Endospore stain	Negative
6.	Motility	Negative
BIOCHEMICAL		
1.	Urease	+
2.	Indole Production	+/–
3.	H_2S in T.S.I.	+/–
4.	Citrate Utilisation	+/–
5.	PPA - Slope Test	+
6.	PPA Liquid test	+
7.	V-P Test	–
8.	Gas production in :	
	a. Glucose	+/–
	b. Lactose	+/–
	c. Mannitol	+/–
	d. Salicin	+/–
9.	Gelatinase	+/–

Discussion

The genus *Proteus* is characterised by morphological properties such as rod shaped cells, presence of flagella, swarming nature, a gram negative response, absence of spores, sheathes, prosthecaes or capsules and by physiological properties like a chemoorganotrophic nature ability to ferment glucose, mannose and maltose but inability to produce acid and gas from lactose. If the above characters are found in an isolate there is every possibility of it being *Proteus* sp.

The properties of the four species *Pr. vulgaris, Pr. mirabilis, Pr. morganii, Pr. rettgeri* are very distinct and the four can be distinguished using simple biochemical tests like Indole production, Urease production, H_2S production, Salicin fermentation etc.

Key Characters of *Proteus* sp.
PPA (+)
Urease +
Pr. vulgaris, Pr. mirabilis, Pr. morganii, Pr. rettgeri

Table 13 Non-Lactose Fermenters

	Indole	Mannitol	H_2S	Salicin	VP	Maltose	Citrate	Gelatin	Glucose
Pr. vulgaris	+	–	+	AG	–	AG	–	+	AG
Pr. mirabilis	–	–	+	AG	+	–	+	+	AG
Pr. morganii	+	–	+	–	–	–	–	–	AG
Pr. rettgeri	+	–	–	A	–	–	+	–	A

A – Acid, AG – Acid and Gas, G – Gas

Pr. mirabilis cannot produce acetyl methyl carbinol and indole, the other 3 can break tryptophan to yield indole. However, H_2S is involved in T.S.I. only by *Pr. vulgaris* and *Pr. morganii* and *Pr. rettgeri* being prominent in the inability to produce H_2S.

Among the remaining *Pr. vulgaris* and *Pr. morganii* the former can ferment salicin but latter does not posses this ability.

According to Bergey's manual the four species with the tube are separated by differences in biochemical reaction, in somatic (O) and flagellar (H) antigens and in susceptibility to bacteriocins and bacteriophage.

11

Salmonella

A member of family Enterobacteriaceae encloses in it the motile, flagellated variants, which are gram negative, nonsporing and devoid of capsules. This pathogen has several serotypes which were given a specific name evoking the role *S. typhi* (agent of typhoid), *S. paratyphii* (para typhoid bacterium), *S. abortuovis* (agent of abortion in sheep) and *S. typhimurium* (agent of disease of mouse).

The principal habitat of *Salmonella* is intestinal tract of man and animals. The biochemical tests are fairly constant in this group unlike *Proteus*. These are non lactose and non sucrose fermenters. Acid and gas is usually produced with a wide range of sugars. Indole is not formed. Gelatin and urea are not decomposed.

Aerobic or facultatively anaerobic by nature, forms an initial smooth colony which changes to rough on prolonged incubation.

Gram negatives non sporing (2.5 μ × 0.5 μ) actively motile peritrichate flagella.

Materials and Methods

Isolation

1. Take about 5 ml of sewage sample.
2. Prepare serial dilutions.
3. Inoculate 0.1 ml of the highest dilution on Mac-Conkey agar and deoxycholate citrate agar plates. (composition given in chapter 10).
4. Incubate at 37°C for 2 days.
5. Colourless, translucent, pale, non lactose fermenting colonies will develop.

Enrichment method

(i) Sample can also be inoculated in Wilson and Blair medium which inhibits growth of *E. coli* while *Salmonella* develops first as greenish colony with black centre which enlarges to fill the entire colony after 24-48 hrs. of incubation. The colonies are round, about 2 mm in diameter black and surrounded by metallic sheen.

(ii) In S-S agar, it grows as a black smooth colony turning to rough on ageing.

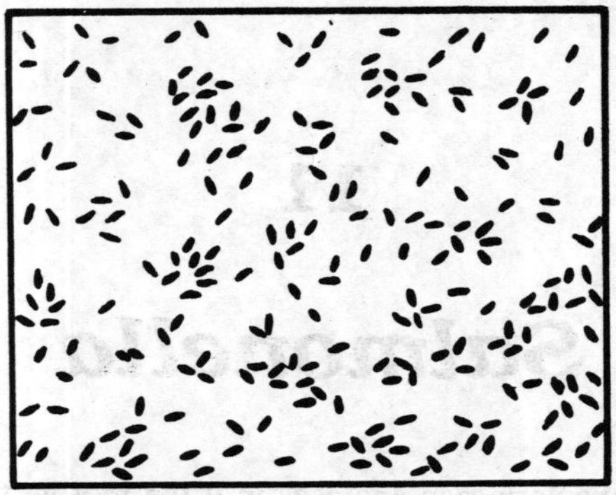

Fig. 11.1 Culture of *Salm. typhosa.*

(1) Composition of Wilson and Blair's Medium

(A) Bismuth sulphite : This is nutrient agar supplemented with (A) and (B) solution. Dissolve 30 g of bismuth-ammonium citrate scales in 100 ml distilled water. To this, add 100 g anhydrous sodium sulphite. Make final volume 1000 ml with distilled water and add 21 g of sodium phosphate crystals ($Na_2HPO_4 . 12H_2O$). Boil and cool. This mixture will keep for months.

(B) Iron citrate brilliant green mixture

Mix 1% solution of ferric citrate with 1% brilliant green and make volume with distilled water to 25 ml. The mixture will keep for months.

To 1 litre of sterilised nutrient agar aseptically add 70 ml of (A) and 4 ml of (B). Mix and plate.

(2) Composition of SS - Agar

Beef extract	5 g
Lactose	10 g
Bile salt mixture	8.5 g
Sodium citrate	8.5 g
Sod. thiosulfate	8.5 g
Ferric citrate	1 g
Agar	13.5 g
pH	7
D.W.	1000 ml

Morphological and Biochemical Tests

S. No.	Tests
A.	MORPHOLOGICAL
1.	Colony
2.	Gram stain
3.	Flagella stain
4.	Capsular stain
5.	Endospore stain
6.	Motility
B.	BIOCHEMICAL
1.	Acid and gas production in glucose broth
2.	Acid and gas production in lactose broth
3.	H_2S from TSI
4.	Acid and gas production in mannitol
5.	Inositol fermentation
6.	Xylose fermentation
7.	Gelatinase
8.	IMViC reactions
9.	Urease

Identification

Identification of genus is quite easy owing its coliform nature supplemented by the to inability to ferment lactose, sucrose, dextrose and manitol motile in nature.

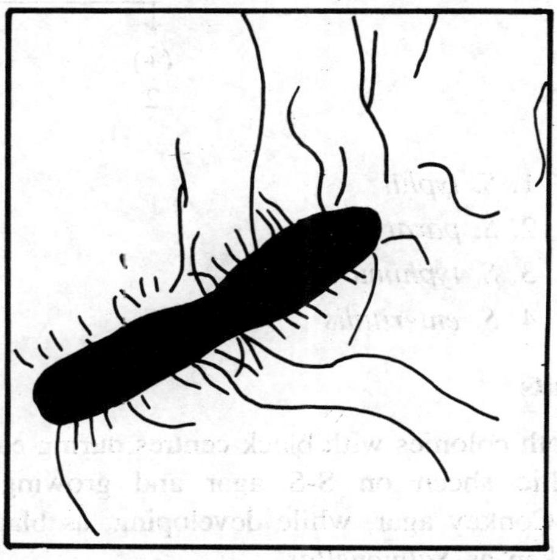

Fig. 11.2 *Salmonella typhii.*

DICHOTOMOUS CHART FOR IDENTIFICATION OF SPECIES

Wilson and Blair's Medium

↓

Black Colonies with melatic sheen

↓

Lactose broth

↓

(–)

Salmonella

↓

Inositol

(+) (1, 2, 4) | (–) (3)

↓

Glucose broth

Mannitol broth

Xylose broth

Acid (1) | Acid and Gas (2, 4)

↓

H_2S in T.S.I.

(+) 2 | (–) 4

1. *S. typhii*
2. *S. paratyphii*
3. *S. typhimurium*
4. *S. enteritidis*

Results

Smooth colonies with black centres during early periods of incubation turning to rough with metallic sheen on S-S agar and growing as white translucent mass of bacteria on Mac-Conkey agar, while developing, as black colony on Wilson and Blair's medium can be taken as *Salmonellas.*

The usual biochemical reactions alone are not sufficient to establish the identity of *Salmonella.* The negative urease reaction, no evolution of gas or acid production in lactose broth however, are the strict *Salmonella* characteristics.

Table 14 Summary of Results (Expected)

S. No.	Tests	Results
A.	MORPHOLOGICAL	
1.	Colony on :	
	a. Mac-Conkey agar	White
	b. SS agar	Black
	c. Wilson and Blairs medium	Black
2.	Gram stain	Negative
3.	Flagella stain	Positive
4.	Capsule stain	Negative
5.	Endospore stain	Negative
6.	Motility	Positive
B.	BIOCHEMICAL	
1.	Glucose Broth	Positive
2.	Mannitol Broth	A/A G
3.	Xylose Broth	A/A G
4.	Inositol Broth	+/–
5.	Lactose Broth	Negative
6.	H_2S from TSI	+/–
7.	Gelatinase	+/–
8.	Urease	–
9.	Indole production	–
10.	Methyl Red Test	+/–
11.	VP-Test	–
12.	Citrate Utilisation	+/–

A – Acid only; AG – Acid and Gas both;
(+) Positive; (–) Negative

Discussion

Peritrichous flagella, absence of spores, prosthecas capsules, gram negative reaction, a chemoorganotrophic mode of nutrition, ability to ferment glucose and mannose. Strictly non lactose fermenter. Above were some of the principle characters of *Salmonella.*

Microscopic observations and the sudden change in colony texture from smooth to rough, with the increasing length of incubation, an obligately respiratory metabolism with oxygen as the terminal electron acceptor do not leave any doubt that the isolate is *Salmonella.*

The establishment of species on the basis of ability to ferment glucose, sucrose, xylose and inositol. The latter is broken down with evolution of acid by *S. typhii, S. paratyphii* and *S. enteritidis* but not by *S. typhimurium* species. A change in the colour of the broth confirming acid production will be seen in case of *S. typhii, S. paratyphii* and *S. enteritidis* but there will be no evolution of gas in *S. typhii.* This leaves with a clear identification character between *S. paratyphii* and *S. enteritidis* which can be done by observing H_2S production in T.S.I. The hydrogen sulfide gas is evolved by *S. paratyphii* but *S. enteritidis* fails to do so.

Table 16 Key Characters of *Salmonella* sps.

Test	*S. typhii*	*S. paratyphii*	*S. typhimurium*	*S. eneteritidis*
Motility	+	+	+	+
Lactose	–	–	–	–
Glucose	A	AG	AG	AG
Mannose	A	AG	AG	AG
Dulcitol	–	AG	AG	AG
Xylose	A	AG	AG	AG
Inositol	–	AG	AG	–
H_2S	+/-	+	+	–
Urease	–	–	–	–

12

Vibrio

Vibrio is fairly large and important group of gram negative, comma shaped rods 1.5 µ to 3 µ in length by 0.2 µ to 0.4 µ in breadth arranged singly or in pairs. It is actually motile, showing darting or scintillating movement with the single large flagellum. In old culture involution forms occur in the form of S or spirillium.

The aerobes can grow on ordinary media and develop into small colonies within 24 hrs. They can ferment a variety of sugars but remain passive towards lactose. The prominent species is *Vibrio cholerae,* the casual organism of cholera. Oxidase positive reaction shows presence of oxygen as an essential factor for final transport of the electron through E.T.C. Indole production and urea hydrolysis are positive reactions however, gelatin liquefaction is not performed. The characters of *Vibrio cholerae* will be taken as key for the genus.

MATERIALS AND METHODS

I. Composition of Medium

(Thiosulfate - citrate - bile salt - sucrose) (TCBS) agar
Kobayashi et al., 1963.

Yeast extract	5 g
Peptone	10 g
Sucrose	20 g
Sodium thiosulfate	10 g
Sodium citrate	3 g
Ox-gall	5 g
Sodium chloride	10 g
Ferric citrate	1 g
Thymol blue	0.04 g
Bromothymol blue	0.04 g
Agar	15 g
D.W.	1000 ml
pH	8.6

The medium is heated to dissolve the ingredients cooled to 45°C and poured into petridishes. It should not be autoclaved.

Direct Isolation

1. Take 1 g of sludge sample.
2. Dissolve it in distilled water and prepare serial dilutions.
3. Spread 0.1 ml of highest dilution on the TCBS agar plate.
4. Observe the development of colonies.

II Reed's modification of Wilson Blair's medium

It is Wilson Blair's medium to which 1% sodium chloride and mannose are added and final pH is adjusted to 9.2. It has the advantage that it inhibits the growth of non-agglutinable *Vibrios*.

Composition of Wilson-Blair's medium is given in chapter 11.

III. Pellicle formation in peptone water

Identification

The identification of *Vibrio* a member of family *Spirillaceae* is accomplished using the usual biochemical reactions.

S. No.	Tests
A	MORPHOLOGICAL
1.	Colony
2.	Gram stain
3.	Flagella stain
4.	Endospore stain
5.	Capsule stain
6.	Motility
B	BIOCHEMICAL
1.	Oxidase test
2.	Indole production
3.	Citrate utilisation
4.	H_2S from T.S.I.
5.	Urease
6.	Gelatinase
7.	Growth in peptone water
	a. +8% NaCl
	b. – NaCl

8.	Growth at 42°C
9.	Fermentation of
	a. Glucose
	b. Lactose
	c. Inositol
	d. Salicin
	e. Sucrose

Results

The translucent, smooth, bluish colonies on nutrient agar developing into a pellicle like growth on peptone water can be taken as an aspirant for the genus *Vibrio*. Each cell bearing a spiral rod shaped structure with a single long flagellum confirms the morphological characters of *Vibrio*. The positive oxidase reaction, indole production, Simmon's citrate utilisation, gelatin hydrolysis, glucose fermentation etc. are some key characters of *Vibrio* group.

Table 16 Summary of Results (Expected)

S. No.	Tests	Result
A	MORPHOLOGICAL	
1.	Colony	Cream coloured
2.	Gram Reaction	–
3.	Flagella stain	+
4.	Endospore stain	–
5.	Capsule stain	–
6.	Motility	+
B	BIOCHEMICAL	
1.	Oxidase Test	+
2.	Indole Production	+
3.	Citrate Utilisation	+ (delayed)
4.	H_2S from T.S.I.	–
5.	Urease	–
6.	Growth in peptone water	Pellicle
	a. +8% NaCl	+
	b. – NaCl	–
7.	Growth at 42°C	+

(Contd.)

8.	Fermentation :	
	a. Glucose	AG
	b. Lactose	–
	c. Inositol	AG
	d. Salicin	–
	e. Sucrose	–
9.	Gelatinase	+

Discussion

The rod shaped cells, with a long single flagellum, thin cell wall, low peptidoglycan content, if observed in an isolate, it should be taken as *Vibrio*.

The positive oxidase reaction confirms oxygen as an ultimate electron acceptor in the E.T.C., the production of H_2S in T.S.I., negative response to the urease and a positive gelatinase reaction.

Inability to ferment inositol and salicin, formation of pellicle in peptone water broth confirms that the isolate is *Vibrio*.

According to Bergey's mannual it was isolated from faeces, the enrichment media used is TCBS agar, gelatin agar and the peptone water broth. The G + C content varies from 40-50 mole %.

13

Xanthomonas

Introduction

These are rod shaped, aerobic, polarly flagellated monotrichous bacteria. They have an antagonistic association with higher plants and develop into yellow coloured mucoid colonies on sugar containing media. This characteristic feature is the unique property of the brominated, aryl-polyene esters (Xanthomonadian) pigments and the cellulosic slimes. Besides these features. *Xanthomonades* are aerobic, oxidative and catalase positive. Intolerant to triphenyl tetrazolium chloride (TTC), capable of utilising narrow selection of C source but incapable of utilising aspargine as sole source of C and N, slow growing, lipolytic, proteolytic, pectiolytic and or amylolytic, NR (–) (NR — nitrate reductase) intolerant to NaCl are some additional reactions of *Xanthomonas*.

Isolation

SX Agar based on Starr's medium

Soluble potato starch	10 g
Beef Extract	1 g
Ammonium chloride	5 g
Potassium diphosphate	2 g
Methyl violet B	1 ml of 1% in 20% ethanol
Methyl green	2 ml of 1% solution
Cycloheximide	250 mg
Agar	15 g
D.W.	1000 ml
pH	6.8

DARK INCUBATION

Direct Isolation

1. Add about a pinch of soil in 5 ml D.W.
2. Make serial dilutions
3. Spread 0.1 ml of highest dilution on SX agar plates

4. Incubate for 3 days at 30°C and observe the colony morphology
5. Starch digestion can be observed after 5 days of growth

Morphological and Biochemical Tests

After 3 days of incubation small transluscent colonies surrounded by clear zone of 4.5 mm diameter digestion as a result of starch with in 5 days. The colonies grow into 3 mm in diameter mucoid, glistening, convex circular with entire margin, translucent with bluish purple centres. The starch digestion zone has a diameter of 8-10 mm.

S. No.	Tests
A	MORPHOLOGICAL
1.	Colony
2.	Gram stain
3.	Endospore stain
4.	Capsule stain
5.	Flagella stain
6.	Motility
B	BIOCHEMICAL
1.	Oxidase
2.	Catalase
3.	Amylase
4.	Caseinase
5.	Pectinase
6.	Nitrate reduction
7.	NaCl tolerance 2%

Results

Xanthomonas are expected to develop into 4-5 mm diameter zone of starch digestion. By 5 days they develop into mucoid colonies with entire margins and translucent with bluish purple centres the starch digestion zone is 8-10 mm in diameter.

It is a gram negative, polarly flagellated with positive catalase and oxidase reactions. Nitrate reduction should be negative.

Discussion

Xanthomonas can be identified by morphological properties such as rod shaped cells, presence of polar flagella a gram negative reaction, slime formation by the unique property of xanthomonadian pigment, brominated, aryl-polyene esters, absence of photosynthesis and inability to fix nitrogen.

Growth on selective SX agar medium and microscopic observations along with positive reactions towards catalase, oxidase, amylase, Absence of colour change in glucose broth with bromocresol purple indicator and durham tube confirms its aerobic nature.

No reduction of nitrate, no evolution of ammonia and nitrogen gas. Positive amylolytic, pecuolytic activity with intolerance to NaCl assign the isolate to the genus *Xanthomonas* with G + C content ranging from 63-71.

Table 17 Summary of Results

S. No.	Tests	Results
A	MORPHOLOGICAL	
1.	Colony	Yellow
2.	Gram Reaction	Negative
3.	Flagella stain	Positive
4.	Capsule stain	Negative
5.	Endospore stain	Negative
6.	Motility Test	Positive
B	BIOCHEMICAL	
1.	Oxidase	Positive
2.	Catalyse	Positive
3.	Amylase	Positive
4.	Caseinase	Positive
5.	Pectinase	Positive
6.	Nitrate reduction	Negative
7.	2% NaCl tolerance	Negative

14

Serratia

Serratia is a member of Enterobacteriacae having synonymus like *Chromobacterium, Micrococcus* found in water, air, bread, milk potatoes etc. They are motile with 2-4 peritrichate flagella. Colonies are 1-2 mm in diameter after a growth of 48 hrs. with low convex smooth, translucent surface and vary from colourless to bright red in colour.

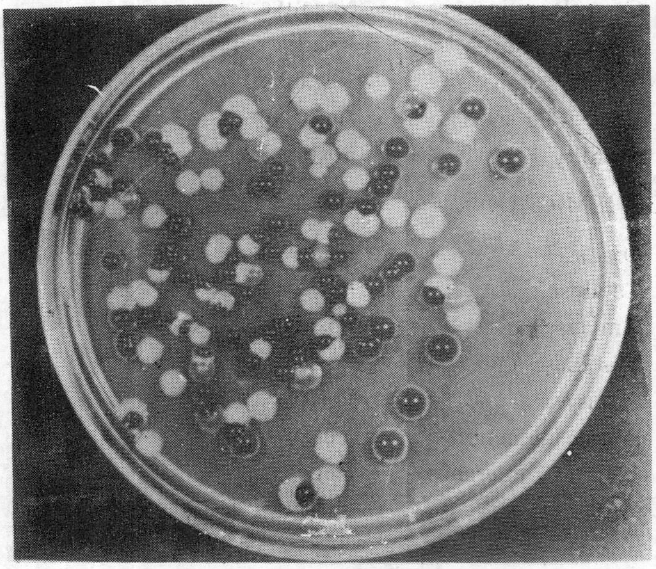

Fig. 14.1 *Serratia* **colonies.**

Some species and biotypes produce a non-diffusible red pigment digiosin which is 2 methyl-3 amyl-methoxy prodigiosene.

Identification

Identification of *genus* is based on usual biochemical tests. Presence of pigments and use of selective media play an important role in this genus.

Five species have been defined, these are *S. marcescens, S. proteamaculans, S. plymutheca, S. marinorubra* and *S. odorifera.*

Each bacterium is a small, gram negative rod, oval or coccobacillary, 0.7 μ to 1 μ by 0.7 μ with rounded ends 2-4 peritrichate flagella the cells are devoid of spores.

Serratia can be identified by usual biochemical coliform reactions with some specific tests like gluconate, iodoacetate, xylosidase tests. Production of extracellular gelatinase, lecithinase and DNase are special characteristics of *Serratia.*

Isolation

Pigmented species give pink to red colonies on nutrient agar. While the non-pigmented ends appear as opaque, whitish, mucoid or transparent, smooth colonies.

The salt tolerance and relatively low growth temperature of all *Serratia* may help to devise a means of enriching media. A nutrient broth containing 4% NaCl and incubated at 15-20°C will allow multiplication of *Serratia* but not of *Enterobacter cloacaceae* or *Aeromonas hydrophilla.*

Capryllate Thallus (CT) Agar medium has been devised for the selective isolation of all *Serratia* species and biotypes. Thus CT agar derived from M-70 minimal medium is made as follows (Starr et al., 1976) :

CT agar for selective isolation of *Serratia*

Preparation of trace element solution

D.W.	1000 ml
H_3PO_4	1.96 g
$FeSO_4 . 4H_2O$	0.0556 g
$MnSO_4 . 4H_2O$	0.0223 g
$CuSO_4 . 5H_2O$	0.025 g
$CO (NO_3)_2 6H_2O$	0.03 g
H_3BO_3	0.0062 g

Store at 4°C.

Solution A

$CaCl_2 \cdot 2H_2O$	0.0147 g
$MgSO_4$	0.123 g
KH_2PO_4	0.0680 g
K_2HPO_4	2.610 g
Trace element solution (see above)	10 ml
Caprylic acid	1.1 ml
Yeast (5% wt/vol.)	2 ml
Thallous sulfate	0.25 g
D.W.	500 ml

Adjust the pH to 7.2 using NaOH. Autoclave at 110°C (or 120°C for 20 min).

Solution B

NaCl	7 g
$(NH_4)_2 SO_4$	1 g
Agar	15 g
D.W.	500 ml

Adjust pH to 7.2 Autoclave and mix solution A and B asceptically and the resulting media is poured in thick layers into sterile plastic petri dishes (25-30 ml for petri dishes 9-10 cm in diameter). The media is no longer effective if it is remelted, but plates of CT agar keeps well for several weeks at 4°C if contamination and desiccation are prevented.

Tests for identification

S. No.	Tests
A	MORPHOLOGICAL
1.	Colony
2.	Flagella
3.	Gram Reaction
4.	Capsule stain
5.	Endospore stain
6.	Motility Test
B	BIOCHEMICAL
1.	Growth with 4% NaCl
2.	Growth at 14°C and 20°C
3.	V.P. Test
4.	Lipase activity
5.	Citrate utilisation
6.	Urease
7.	Amylase

For the identification of species some specific tests should be performed. These are :

(i) Growth on 5 mg/ml Tetra cycline (Tc)

(ii) Tetrathionate reduction

Add 5 mg/ml of Tc to nutrient agar. The tetrathionate reduction can be accomplished as follows :

Tetrathionate Reductase

The liquid medium [Le Minoreta (1970)] is composed of :

Potassium tetrathionate			5 g
Bromothymol blue (0.2% aq solution)			25 ml
Peptone water :			
Peptone	10 g	q.s.	1000 ml
NaCl	5 g		
D.W.	1000 ml		

Adjust pH to 7.4. Filter, sterlise and dispense into small durham tubes. Inoculate and observe after 1-2 days.

C source utilisation tests

In the C.T. Agar medium by omitting the yeast extract, thallus sulfate and caprylate from *solution A*. Neutralised C sources are added in solution A to give a 0.1% final concentration in the complete medium 0.2% in the case of carbohydrates.

Specific Tests

a. Gas production in glucose agar

Glucose agar consists of nutrient agar supplemented with 1% glucose. Prepare and dispense into tubes. Autoclave. Cool and inoculate before the agar sets. Examine the tubes for bubbles of gas upto 3 days.

Incubated for 2 days and the reaction is considered positive when a light green to rusty-yellow colour develops. A deep blue colour is given by a negative test.

b. Iodoacetate test [Lysenko (1961)]

Tryptone	1 g
Yeast extract	1 g
NaCl	5 g
K_2HPO_4	0.3 g
Glucose	10 g
Bromothymol blue 1.5% aq solution	2 ml
D.W.	1000 ml
pH	7.2

Above medium is sterilised by boiling for 30 minutes on each of 3 successive days. Sodium iodoacetate (filter sterilised) is added to a final conc. of 10^{-3} M (208 mg/l). The inoculated medium is examined for acid production each day for 5 days. Controls without iodoacetate are also inoculated.

c. β-*Xylosidase* [Brison, Richard and Leariot (1972)]

Aqueous (1% wt/vol) p nitrophenyl β-xyloside is dispensed into the wells in 0.05 ml amounts followed by 0.05 ml of a fresh bacterial suspension in 0.25 M phosphate buffer pH 7 Examine the plates after 24 hours.

d. Susceptibility to 5-Fluoro cytosine

For determining susceptibility to 5-FC an M-70 minimal agar plate containing 0.2 (wt/vol) glucose is inoculated with few drops of culture in broth spread on the plate. After drying for 15 minutes at 17°C, a paper disc impregnated with 100 mg 5-FC is placed on the plate. After over night incubation, the plate is examined for an inhibition zone.

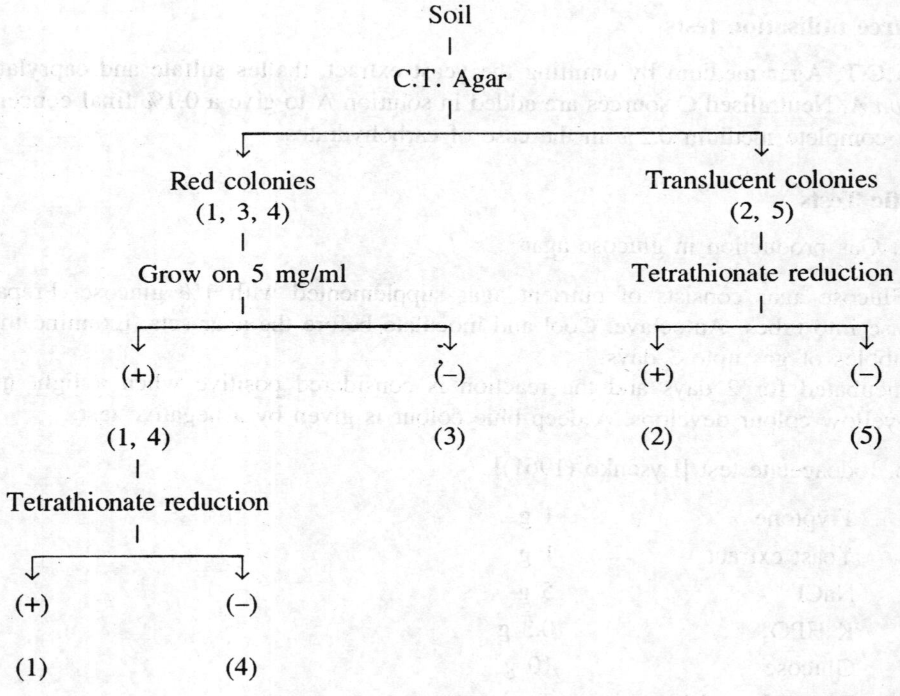

1. *S. marcesens*
2. *S. proteamaculans*
3. *S. plymuthica*
4. *S. marinorubra*
5. *S. odorifera*

Results

An aerobically growing *Serratia* produces red or white translucent colonies on C.T. agar depending on the species. It can tolerate 4% NaCl solution and is also able to grow at low temperatures. Citrate utilisation and Voges-Proskauer's tests are positive. It can produce acid from glucose and maltose and can also hydrolyse lipid and protein.

Besides xylosidase activity, iodoacetate, gluconate, tetrathionate reductase tests are also positive.

The species can be identified using the following flowsheet.

Table 18 A list of summarised results of the various biochemical tests is as follows :

	1	2	3	4	5
1. Red pigment	+	–	+	+	–
2. Musty vegetable odour	+	–	–	+	–
3 Growth at 5°C	–	+	+	–	+
4. Growth at 40°C	+	–	–	+	–
5. Growth with KCN	+	+	–	+	+
6. Indole production	–	–	–	–	+
7. Tetra thionate reduction	+	+	–	–	–
8. Gas from glucose agar	–	+	+	–	–
9. Acid from sucrose	+	+	+	+	+
10. Chitin hydrolysis	+	+	+	+	–
11. Tween - 80 hydrolysis	+	+	+	+	–
12. Growth with 5 mg/ml Tc	–	–	–	–	–
C. SOURCE UTILISATION					
13. Cellobiose	+	+	+	+	–
14. Lactose	+	+	+	+	–
15. Sucrose	m	+	m	+	m
16. H_2S	–	–	–	–	–
17. V.P. Test	+	–	+	–	–
18. Gelatinase	+	+	+	+	+
19. Amylase	+	+	+	+	+
20. Growth with 4% NaCl	+	+	+	+	+
21. Citrate utilisation	+	+	+	+	+

Discussion

The genus *Serratia* is characterised by morphological properties such as the rod-shaped cells which are either straight or slightly curved in one plane, the presence or absence of spores, sheaths or prosthecas a chemo-organotroph with respiratory metabolism an absence of photosynthesis and fermentation and ability to utilise various C sources. Very few coliforms are pigmented and if a pigmented isolate can grow on C.T. agar with salt tolerance, low temperature and ability to hydrolyse protein, starch and lipid but lacks the potential to use urea assigns the bacterium genus *Serratia.*

Moreover, a positive test with 5 mg/ml tetra cycline tetrathionate reduction, accompanied with utilisation of range of C compounds, positive gluconate, iodoacetate and β-xylosidose leaves no doubt of the isolate being *Serratia.*

The species can be confirmed through gas production in glucose and tetrathionate reduction as a promising reaction without ambiguity.

The results if properly fitted to the key characters place the isolate in the genus *Serratia* according to Bergey's manual. It has G + C content of 40-53 mole %.

15

Staphylococci

The genus *Staphyloccous* is a group of gram positive, spherical bacteria that forms irregular cellular aggregation or grape like clusters but seldom, form chains when grown in liquid medium. They produce characteristic pigment ranging in colour from orange to golden yellow to pink to white. All pathogenic strains are coagulase positive while non pathogenic species are not. They are also β-haemolytic. *S. aureus* is pathogenic. In human beings disorders caused by pathogenic *Staphylococci* are cellulitis, pustules, boils, carbuncles, impetigo and secondary infections. One of the most common type of food poisoning is *Staphylococcal* food poisoning caused by enterotoxin excreted in food.

Among the most serious staphylococcal infections are septicemia, meningitis, pneumonia and osteomyelitis.

Staphylococci can be differentiated from *Micrococci* on the basis of anaerobic fermentation of glucose. All pathogenic species are mannitol fermenters.

Staphylococcus Isolation

These are found on skin and mucous membranes of warm blooded animals and in some food and curing brines.

A simplest enrichment procedure for *Staphylococcus* is to culture the source material anaerobically in a medium containing 7% NaCl. The high osmotic concentration of this medium inhibits osmotically sensitive organisms. The culture is purified by streak plating from the enrichment broth and selecting a colony composed of organisms that are catalase positive, gram positive cocci and ferment glucose anaerobically. It produces golden yellow to orange pigment on plates.

MATERIALS AND METHODS

Enrichment

(i) Inoculate the source material in 7% NaCl broth and seal the tube with vaspar or vaseline.

(ii) Incubate at 30°C for 2 days.

(iii) Streak the culture from enrichment broth onto the plate count agar plate and incubate at 30°C for 2 days.

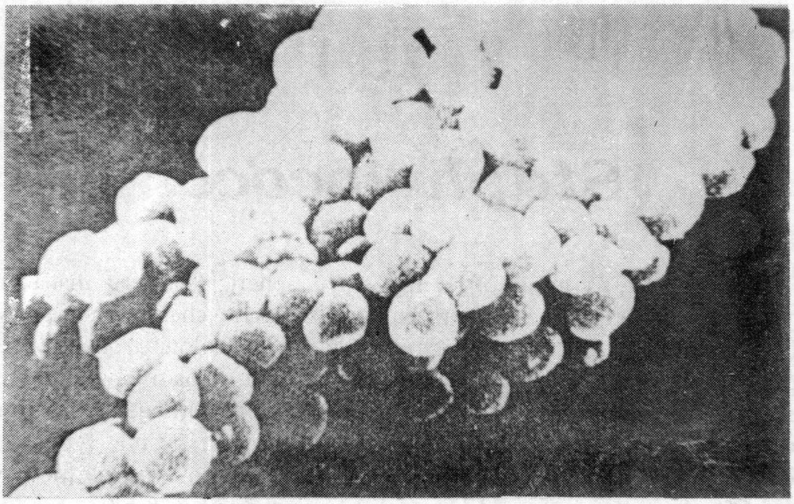

Fig. 15.1 *Staphylococci* (Electron micrograph).

Plate count agar

Tryptone	5 g
Yeast extract	2.5 g
Glucose	1 g
Agar	15 g
D.W.	1000 ml

Isolation

1. Examine the isolated colonies on streak plate for catalase test
2. Prepare a wet mount for the study of the morphology of the bacterium
3. Emulsify a small bit of colony in sterile broth and streak on a plate count agar plate. Incubate both at 30°C for 24 hrs.

Confirmation

1. Examine the streak plate for purity to assure that a single colony type was transferred.
2. Prepare a Gram stain.
3. Inoculate a bromocresol purple glucose broth from the overnight broth culture and incubate at 30°C for 5 days.

Tests for Identification

S. No.	Tests
A.	MORPHOLOGICAL
1.	Colony
2.	Flagellar stain
3.	Capsular stain
4.	Endospore stain
5.	Gram stain
B.	BIOCHEMICAL
1.	Gas production from glucose and mannitol
2.	Gelatinase
3.	Haemolysis
4.	Coagulase
5.	DNAse
6.	Novobiosin sensitivity test
7.	Indole production

SPECIFIC BIOCHEMICAL TESTS

1. Coagulase activity

Principle : Coagulase positive *Staphylococci* form an enzyme coagulase, that forms a clot from normal blood plasma clot formation constitutes a positive test.

The blood used is prevented from clotting by addition of citrate or oxalate or versene buffers, and the blood cells in suspension are centrifuged out. The plasma is employed in test.

(i) Add 0.5 ml of versenated plasma to test tube.

(ii) Now add enough cells of the exponential culture under study to give a turbid suspension.

(iii) Incubate at 37°C for 1 hr and examine.

2. DNase Test

In addition to coagulase reaction most infectious *Staphylococcus* produce a thermostable enzyme nucleus that deploymerise DNA. This enzyme, the DNase, can be tested using agar containing dissolved DNA and a dye, toluidine blue O which is exposed to a heated culture of organism. Presence of nuclease develops pink zones where DNA has been hydrolysed.

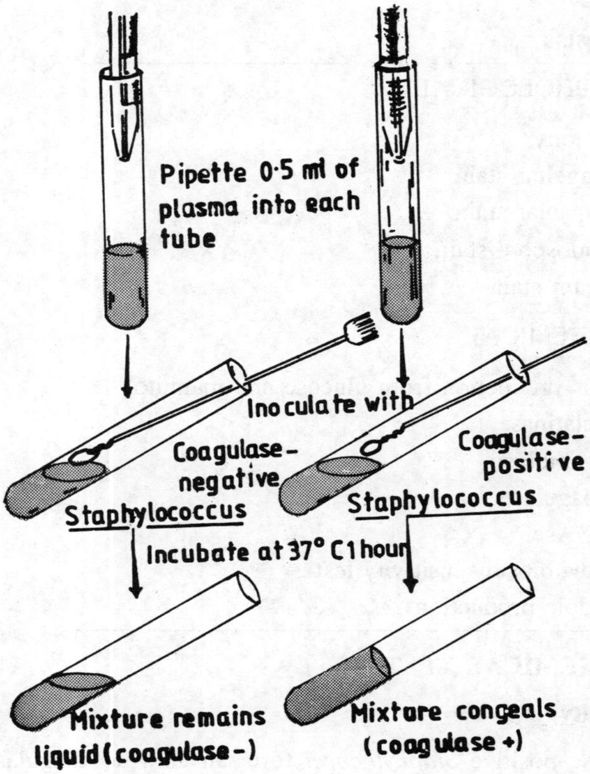

Fig. 15.2 Steps in the coagulase test.

Composition of Toluidine blue DNA Agar – (TDA)

DNA	0.3 g
$CaCl_2$ (0.01 M)	1 ml
NaCl	1 g
Agar	10 g in 0.05 M Tris buffer pH 9.
D.W.	1000 ml

After boiling the mixture to completely dissolve the DNA and agar, add 3 ml of 0.1 M toluidine blue O.

Method

1. Carefully pipette 3 ml of melted TDA agar onto a slide and allow it to harden.
2. Remove two small plugs to form equally spaced wells in the hardened agar.

3. Place tubes of *Staphylococcus* culture in a boiling water bath for 15 minutes and cool.
4. Leave one set of cultures unboiled.
5. Transfer a loopful of unboiled *Staphylococcus* culture to well No. 1 and the boiled culture to well No. 2.
6. Incubate slides at 37°C for 2 hours in a moist chamber.
7. Watch for pink rings.

3. Novobiocin Sensitivity

Composition of glucose nitrate salt agar

Glucose	10 g
$NaNO_3$	10 g
K_2HPO_4	10 g
KCl	10 g
$MgSO_4 \cdot 7H_2O$	10 g
Agar	15 g
D.W.	1000 ml

1. Pour three plates of glucose nitrate salt agar
2. Surface plate 1×10^{-3}, 1×10^{-4}, and 1×10^{-5} dilutions of a soil sample and incubate the plate at 30°C for 2-4 days.
3. Inoculate a loopful of above culture in a 3 ml of melted and tempered nutrient agar.
4. Pour the agar on a blank nutrient agar plate and place a novobiocin antibiotic disc on it. Incubate at 37°C.
5. Observe for the clear zones.

4. Mannitol fermentation

(See Chapter 5)

5. Haemolysis

Composition of Blood agar

Tryptone	10 g
Beef extract	3 g
NaCl	5 g
Agar	15 g
D.W.	1000 ml

Table 19 Summary of Results (Expected)

S. No.	Tests	Results
A	MORPHOLOGICAL	
1.	Colony	Circular, pinheaded, convex, entire, translucent colourless on nutrient agar plate
2.	Flagella stain	Negative
3.	Capsular stain	Negative
4.	Endospore stain	Negative
5.	Gram staining	Positive
B	BIOCHEMICAL	
1.	Gas production from glucose and mannitol	Positive
2.	Gelatinase	Positive
3.	Haemolysis	Positive
4.	Coagulase	Positive/Negative
5.	DNAse	Positive
6.	Novobiocin (Sensitivity)	Positive
7.	Indole Production	Negative

Identification of Species

Staphylococcus

- Coagulase negative
  e.g., *Staph. epidermidis*
  Var. *albus*
  Var. *citreus*
- Coagulase positive
  e.g., *Staph pyogenes*
  Var. *aureus*
  Var. *citreus* (rarely)

Sterlise, after cooling to 45°C add 5% fresh defibrinated blood.

(i) Melt a tube (15 ml) of blood base agar and maintain it at 45°C.

(ii) Temper a tube of sterile blood in water bath at 45°C.

(iii) Transfer aseptically 1 ml of blood to the cooled tube of agar and mix with the pipette.

(iv) Mix by rotating the plate, avoiding any bubble formation.

(v) Streak a 24 hour culture of *Staphylococcus.*

(vi) Incubate at 31°C for 24 hrs.

(vii) Examine for hemolysis.

DISCUSSION

The genus *Staphylococcus* is characterised by spherical cells or cocci, absence of flagella, a gram positive reaction, absence of spores, sheaths and by physiological properties like chemoorganotrophic mode, a metabolism dependent on respiration absence of photosynthesis and inability to fix dinitrogen. The isolate showing above characters can be taken as a candidate for the genus *Staphylococcus.*

A positive coagulase reaction, supported by synthesis of thermostable nuclease haemolysis, gas production using glucose and mannitol as substrate all point towards the presence of *Staphylococcus* in the sample. The identification of the species of *Staphylococcus* is based primarily on the coagulase test. A positive coagulase reaction is shown by *S. epidermidis,* while a negative attitude is maintained by *S. pyogens.*

16

Streptococci

The genus *Streptococcus* is characterised by spherical or ovoid cells; which form chains. These are non motile and nonsporing. The growth is scanty in absence of native protein in the media. These produce characteristic hemolytic changes in blood. Only acid is produced by carbohydrate fermentation. *Streptococcus* in general is responsible for sore throat, scarlet fever, impetigo (bacterial endocarditis), rheumatic fever and glomeralonephritis. *S. pyogenes* is also called pathogenic hemolytic *Streptococci* of lancefield group A and the most common infective syndrome is an acutely inflammed throat with or without exudate called tonsilitis. There is acute inflammation often with oedema and fever, wound, burns, chronic skin lesions may become infected with *S. pyogens*, which may extend and cause cellulitis or get into blood stream.

MATERIALS AND METHODS

Isolation

The best source of material for the isolation of *Staphylococcus* is pus sample.

Plate the sample on superimposed blood agar. Pick up small discrete colonies surrounded by clear zone of hemolysis.

Identification

Identification is based on following tests :

S. No.	Test
A	MORPHOLOGICAL TESTS
1.	Colony
2.	Gram staining
3.	Flagella stain
4.	Capsule stain
5.	Endospore stain
6.	Motility

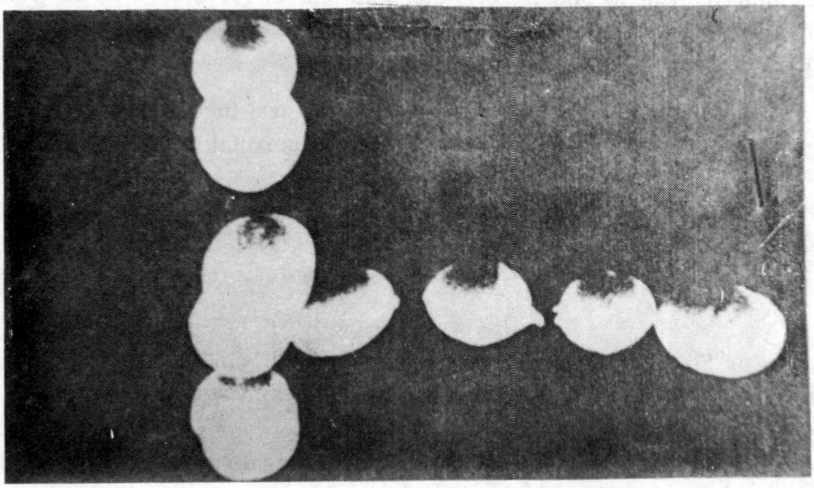

Fig. 16.1 *Streptococci* (Electron micrograph).

B BIOCHEMICAL

1. Glucose broth
2. Nitrate reduction
3. Hemolysis
4. Mannitol fermentation
5. Bile solubility
6. Methyl red test
7. Voges proskauer test
8. Catalase test

Bile solubility test

1. Grow the isolate in meat infusion broth.

Meat infusion broth

Composition

Lean meat	500 g
Peptone	10 g
NaCl	5 g
pH	7.4
D.W.	1000 ml

Add minced meat to 500 ml of water and extract for 24 hrs in the cold. Strain and boil filtrate at 100°C for 15 mins filter. Add peptone and NaCl and make up the volume to 1 litre. Adjust pH to 7.4.

2. Take about 1 ml broth in each of the two test tubes.
3. To one add (0.1 ml) of sodium taurocholate solution (0.5%) and to the other add 0.1 ml of saline.
4. Incubate at 37°C for 30 min and observe for lysis.
5. Confirm by negative staining.

Results

If the isolate is *Streptococci*, it should develop as small, smooth margined, circular, colony semi-translucent with a low convex surface and glistening appearance.

It should produce acid from glucose besides a negative response towards methyl red, bile solubility test, catalase and voges proskauer test.

Table 20 Summarised Results (expected)

S. No.	Tests	Results
A	MORPHOLOGICAL	
1.	Colony	Cream coloured
2.	Flagella	Absent
3.	Endospore	Absent
4.	Capsule	Absent
5.	Motility	Absent
6.	Gram reaction	Positive
B	BIOCHEMICAL	
1.	Catalase	Negative
2.	Glucose broth	Acid
3.	Mannitol broth	Acid
4.	Nitrate reduction	Positive
5.	Hemolysis	Positive
6.	Bile solubility	Negative
7.	Methyl red test	Negative
8.	Voges proskauer test	Negative

Discussion

A circular, cream coloured, semi-translucent colony with entire margin, each cell showing spherical shape, gram positive nature, absence of flagella, capsule and spore is a promising candidate for the genus *Streptococcus*.

The isolate should show a negative catalase reaction, acid production in glucose and mannitol and a positive nitrate test. A strict negative test should be observed while using the isolate in the bile solubility, methyl red and voges proskauer test.

17

Bacillus

Introduction

Rod shaped, spore bearing, usually gram positive, catalase producing and capable of sporulating. Smith *et al.* (1952) sub-divided the genus into 3 groups on the basis of spore shape and sporangial swelling :

Gram positive to gram variable rods, sporulating aerobically and producing catalase

- Spores oval or cylindrical
  - Sporangia not definitely swollen; spore wall thin; Group I *B. megaterium, B. subtilis, B. coagulans*
  - Spores usually oval; sporangia clearly swollen; spore wall thick; Group II *B. circulans*
- Spores spherical; sporangia swollen; non-fermentative casein and starch not hydrolysed; nutritional requirement is complex; Group III *B. calidolactis*

The identification and isolation of *B. megaterium, B. subtilis, B. coagulans, B. circulans, B. calidolactis* will be discussed here since the above species are strictly soil inhabitants.

MATERIALS AND METHODS

Composition of nutrient broth

Beef extract	3 g
Peptone	5 g
NaCl	3 g

Agar	15 g
D.W.	1000 ml
pH	7.2

Direct Isolation

1. Inoculate 1 g of soil in 10 ml of distilled water.
2. Prepare serial dilutions.
3. Spread 0.1 ml of highest dilution on nutrient agar plates.
4. Incubate at 37°C for 2 days.
5. Restreak the isolated colonies on nutrient agar plates.

Selective Isolation

Composition of soil extract agar (Gorden and Regnear, 1963) : To 400 g air dried, sieved garden soil rich in organic matter in a 2 litre conical flask, add 960 ml of tap water. Autoclave at 121°C for 60 min and allow to cool in the unopened autoclave overnight. Then filter the extract through paper and reautoclave the filtrate at 121°C for 20 minutes, and allow to stand at room temperature for at least 2 weeks. The clear supernatant can be used in making soil extract agar.

Peptone	5 g
Beef extract	3 g

Fig. 17.1 *B. rhizoides.*

Agar	15 g
Tap water	750 ml
Soil extract	250 ml

With new batches of soil try small lots of medium between 10 and 100% v/v amount of soil extract. The concentration producing best sporulation should be used for the bulk medium once isolated, *Bacillus* can be grown routinely on nutrient agar medium.

Spread 0.1 ml of the highest dilution on soil extract agar plates.

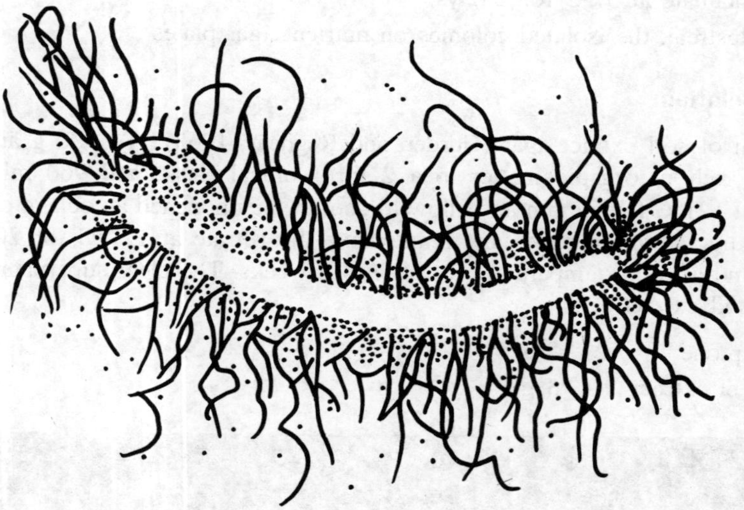

Fig. 17.2 *Bacillus subtilis.*

Identification

Following tests should be performed for identifying the isolate.

S. No.	Tests
A	MORPHOLOGICAL
1.	Colony
2.	Flagella stain
3.	Capsular stain
4.	Endospore stain
5.	Gram stain
6.	Motility test

B BIOCHEMICAL

1. Amylase
2. Gelatinase
3. Caseinase
4. Nitrate reduction
5. Litmus milk
6. Acid form
   a. Arabinose
   b. Mannitol
   c. Lactose.
   d. Xylose
   e. Cellobiose
7. Catalase

Results

The most important identifying character of *Bacillus* is presence of *endospore*. Besides this, these are gram positive and catalase producing.

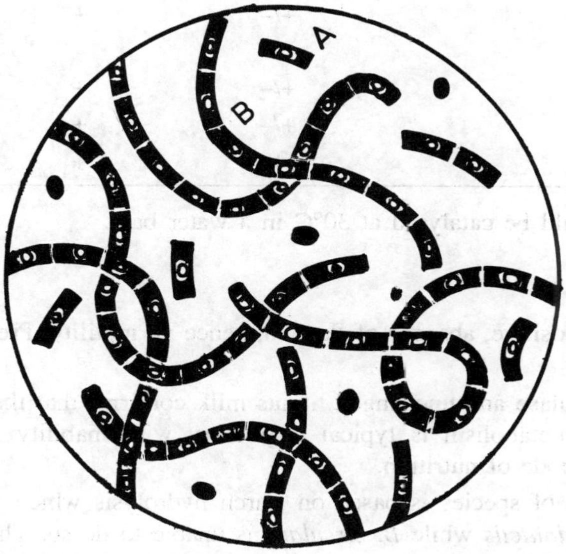

Fig. 17.3 (A) A single bacterium with its *endospore*, (B) A chain of bacteria with their central *endospore*.

Summary of Results (Expected)

S. No.		Group I	Group II	Group III
A	MORPHOLOGICAL			
1.	Colony	Medium	Medium	Medium
2.	Gram stain	+	+	+
3.	Flagella stain	–	–	–
4.	Capsule stain	–	–	–
5.	Endospore stain	+	+	+
6.	Motility	–	–	–
B	BIOCHEMICAL REACTIONS			
1.	Amylase	+	–	+
2.	Gelatinase	+/–	–	+/–
3.	Caseinase	–	–	–
4.	Nitrate reduction	+	–	–
5.	Litmus milk	–	–	–
6.	Acid from			
	a. Arabinose	+/–	–	–
	b. Mannitol	+/–	+	–
	c. Lactose	–	–	+/–
	d. Xylose	+/–	+	+/–
	e. Cellobiose	+/–	+	+/–
7.	Catalase	+	+	+

Note : The tests should be catalysed at 30°C in a water bath.

Discussion

Medium sized, gram positive, absence of flagella, hence no motility. Presence of endospore (centrally placed).

Production of catalase and unchanged litmus milk confirms that the isolate belongs to genus *Bacillus*. The metabolism is typical respiratory with inability to photosynthesize, chemoorganotrophic mode of nutrition.

The identification of species is based on starch hydrolysis which is accomplished by *B. subtilis* and *B. calidolactis* while *B. circulans* is unable to do so. The casein hydrolysis differentiates between *B. calidolactis* and *B. subtilis*. This is further confirmed by acid production in Arabinose.

(DICHOTOMOUS CHART)

Endospore (+)
↓
Litmus Milk *Bacillus*
↓
Amylase

(+)	(–)
(1, 3)	(2)

(1, 3)
↓
Caseinase

(+)	+/–
(1)	3

(1)
↓
Acid in arabinose
(–)
3

1. *B. subtilis*
2. *B. circulans*
3. *B. calidolactis*

18

Lactobacillus

Lactobacillus is a very large group of homofermentative as well as heterofermentative, gram positive bacteria which appears to come from two very distinct type of nutrient habitat.

(i) Man and Animals

(ii) Plant materials

In man and animals they dominate the flora, in the upper gastro-intestinal tract and as in the oral cavity and lower intestine they may form only a minority of the total, strictly anaerobic flora. On plant materials *Lactobacilli* occur in very small number being dominated by the more successful leuconostocs.

The basic morphological characteristics are the rod-shaped usually non motile non spore forming. It produces lactic acid lowering final pH of the medium to 2.2-3.4, resistant to acids and is microaerophillic. They grow well in milk. The identification of genus is based on usual morphological and biochemical tests such as Gram reaction, motility, presence of flagella, oxidase test, catalase reaction, presence of extracellular gelatinase and nitrate reduction.

MATERIALS AND METHODS

Isolation

SL-Medium Composition (Rogosa, Mitchelie and Weismann, 1951)

Trypticase	10 g
Yeast extract	5 g
KH_2PO_4	6 g
Diammonium citrate	2 g
$MgSO_4 \cdot 7H_2O$	0.58 g
$MnSO_4 \cdot 4H_2O$	0.28 g
Glucose	10 g
Arabinose	5 g
Sucrose	5 g
Tween-80	1 ml

Sodium acetate $5H_2O$	2.5 g
Glacial acetic acid 99.5% required to titrate pH of medium to 5.4	
Agar	15 g
D.W.	1000 ml

Dissolve the agar separately by steaming in 500 ml D.W. dissolve all the other ingredients except acetate and acetic acid in 300 ml D.W. without heating, add this to melted agar and steam for further 5 minutes excessive heating must be avoided. Dissolve the sodium acetate in about 15 ml D.W. without heating add acetic acid as a 10% v/v aqueous solution to adjust pH to 5.4. Make up volume to 20 ml. Add this buffer mixture to hot based medium in convenient amount in sterilised screw capped bottles. The medium should be clear light straw coloured giving a firm gel. Avoid repeated melting and cooling. The addition of arabinose and sucrose as well as glucose allows the growth of strains that preferably ferment these sugars.

1. Take about 5 ml of curd and make a suspension in 5 ml of distilled water.
2. Prepare serial dilutions.
3. Plate 0.1 ml of highest dilution on SL agar plates.
4. Incubate at 37°C for 2 days.
5. Observe the colonies.
6. Restreak on to nutrient agar plates to check contamination.

Identification Tests

S. No.	Tests
A	MORPHOLOGICAL
1.	Colony
2.	Gram stain
3.	Flagella stain
4.	Capsule stain
5.	Endospore stain
6.	Motility
B	BIOCHEMICAL
1.	Oxidase
2.	Catalase
3.	Glucose Broth
4.	Gelatinase

5. T.S.I. reactions :
   a. H_2S production
   b. Glucose, lactose and sucrose fermentation in T.S.I.
   c. Acid production
6. Nitrate reduction
7. Litmus milk test
8. Growth at 45°C
9. Cellobiose fermentation
10. Ribose fermentation

Isolate
↓
Cellobiose fermentation

- (+) (1, 2)*
  - Growth at 45°C
    - (+) → (1)
    - (–) → (2)
- (–) (3, 4, 5)
  - Growth at 45°C
    - (–) (3, 4, 5)
      - Ribose fermentation
        - (+) → (3)
        - (–) (4, 5)
          - Lactose fermentation
            - (+) → (5)
            - (–) → (4)

* 1. *L. casei subsp. casei*
2. *L. casei subsp.*
3. *L. curvatus*
4. *L. coryniformis*
5. *L. brevis*

Results

The isolate should grow as a small, smooth margined, white or off white on S.L. medium. Each member is gram positive, immotile rod, without flagella, or spore or sheathe. Catalase and oxidase reaction, should be positive, the confirmation for axenic culture should be made by streaking the isolate on to the nutrient agar plate.

Table 21 Summarised Results

S. No.	Tests	Results
A	MORPHOLOGICAL	
1.	Colony	White
2.	Flagella	–
3.	Gram Reaction	+
4.	Capsule stain	–
5.	Endospore stain	–
6.	Motility	–
B	BIOCHEMICAL	
1.	Oxidase	+
2.	Catalase	+
3.	Glucose Broth	+
4.	Gelatinase	+
5.	T.S.I. Reactions	
	a. H_2S production	+
	b. Glucose, Lactose and Sucrose	+
	c. Acid production	+
6.	Nitrate production	+
7.	Litmus milk	+
8.	Growth at 45°C	+/–
9.	Cellubiose fermentation	+/–
10.	Ribose fermentation	+/–

Discussion

The genus *Lactobacillus* is characterised by morphological properties such as the rod shaped cells, absence of flagella, a gram positive reaction, absence of spores, sheathes, capsules, and by physiological properties such as a fermentative mode of nutrition, absence of photosynthesis and nitrogen fixing ability a potent producer of lactic acid. The isolate from

Fig. 18.1 Culture of *Lactobacilli*.

S.L. Medium bearing above characters can never be any bacteria other than *Lactobacillus*.

The change in the colour of glucose, lactose or maltose broth accompanied by down ward displacement of the liquid due to the production of gas are the promising features of *Lactobacillus* inability to reduce nitrate and a positive oxidase test shows that the bacteria required oxygen as a terminal electron acceptor. Among the four sp. *L. casei. rhamnosus*, the distinction is quite enormous. Since both the subsp. of *L. casei* can ferment cellobiose but the remaining 3 sp. are unable to do so. Among *L. casei* sub-sp. *casei* and sub-sp. *rhamnosus* former can grow at 45°C while latter is unable to do so. The distinction between *L. curvatus* and *L. coryniformis* is made good owing to the inability of *L. curvatus* and *L. coryniformis* to grow at 45°C. Among *L. curvatus* and *L. cornyniformis* former can produce acid and gas from ribose while the latter has strict negative attitude towards ribose.

According to Bergey's manual *Lactobacillus* were isolated from milk. Rogosa (1970) has divided the genus into 3 groups.

Thermobacterium, *Streptobacterium* and *Betabacterium* on the basis of optimal growth temperatures and germentative end products.

Thermobacterium and *Streptobacterium* are homofermentative, the difference lies in the ability of former to grow at 45°C and inability of the latter to do so. All the species discussed here except *L. brevis* belong to *Streptobacterium*. *L. brevis* falls under *Betabacterium*. It is a heterofermentative and requires temperature less than 45°C. The G + C content varies from 30-40 mole %.

19

Membrane Filter Technique

OUTLINE

• Total coliform test • Faecal coliform test • Faecal *Streptococcus* test • Air analysis

A direct and quicker method, it enables to filter large volumes of sample easily. A measured amount of the sample is filtered through a thin bacteria-retaining membrane with a pore size of about 0.45 μm, trapping the bacterial cells on the surface. The membrane is then transferred onto a thin absorbent pad that has been saturated with the particular medium designed to grow or permit differentiation of the organisms sought — a modified Endo medium, for example, if total coliform organisms are sought.

Procedure

1. Attach the filter holder to the rubber stopper, insert into the vacuum flask, and connect the flask to the vacuum line or aspirator.
2. Transfer a sterilised membrane filter to the plateform base of the filter unit place the filter with ruled side up. Use a sterilised forcep to handle the filter.
3. The matched funnel unit should now be placed over the filter disc, while doing so, be sure that it is clamped in place firmly by scissor type clamping.
4. Arrange three petriplates labelled with the sample volumes to be used.
5. Add 20 ml of sterilised buffered water into the funnel before adding the sample.
6. Shake the sample bottle vigorously.
7. Measure out the volume to be sampled with a 10 ml pipette and introduce it into the funnel.
8. Add sterilised buffered water equal in volume to the sample to the funnel twice to rinse cells from the measuring vessel.
9. Turn on the vacuum motor or water aspirator and allow all the liquid o pass through the filter into the flask.

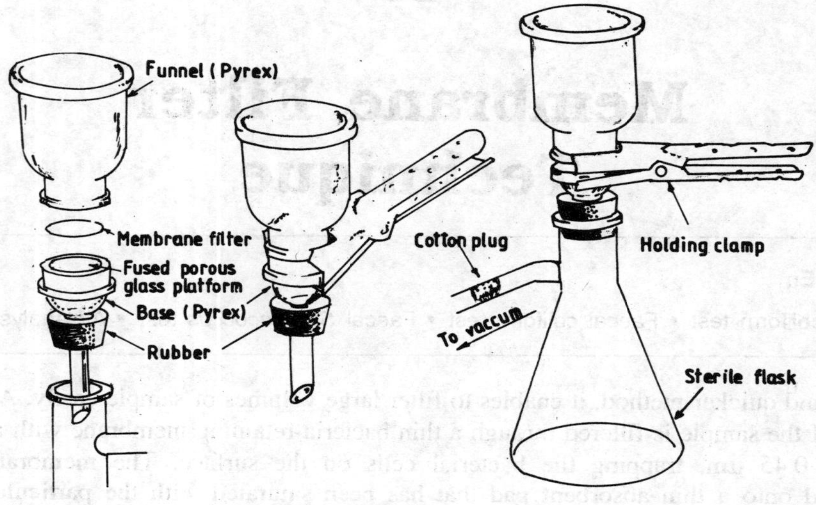

Fig. 19.1 The membrane-filter apparatus.

10. Leave the vacuum on while pouring into the funnel a rinse volume of sterilised buffered water equal to the amount of liquid filtered. Pour this rinse water inside the wall of the funnel leading to a swirling wash.
11. Rinse once again and allow the filter to dry for one minute.
12. Turn off the vacuum source and remove the membrane with flamed sterilised forceps to the desired medium in a petridish.
13. Push the membrane against the far side of the plate and roll it onto the medium to avoid entrapment of air bubbles under the membrane.
14. Incubate all the sets at 27°C.
15. Calculate the number of organisms per 100 ml using the formula.

$$\text{Indicator count per 100 ml} = \frac{100 \times \text{Total No. of colonies counted}}{\text{No. of ml of sample tested}}$$

Total Coliform Test

1. Insert a sterilised absorbent pad into each of three petridishes in a sterilised atmosphere.
2. Add 1.8 to 2 ml of M-Endo broth to the surface of each pad.

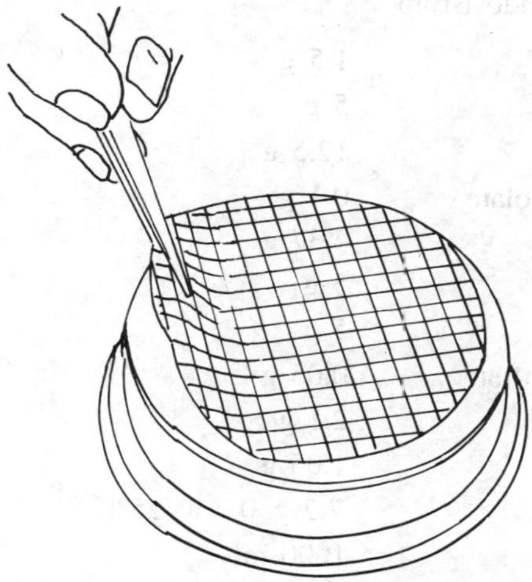

Removal of filter from the filtering apparatus.

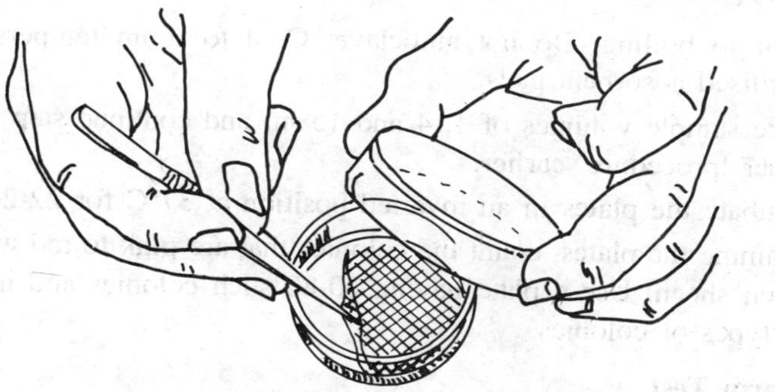

Placing the filter on a pad in a petri plate.

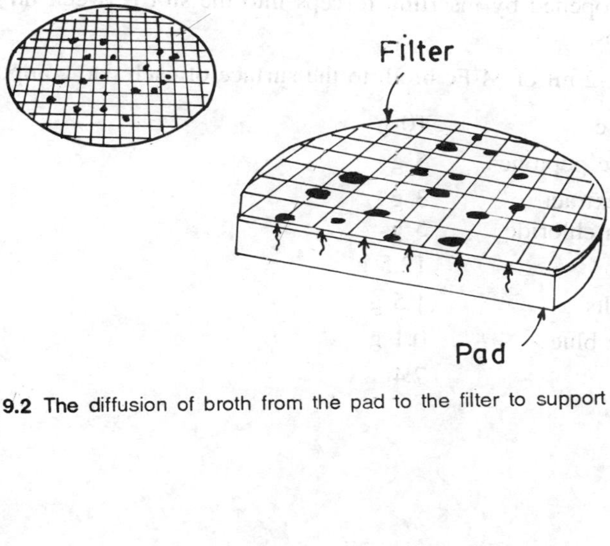

Fig. 19.2 The diffusion of broth from the pad to the filter to support growth.

Composition of M-Endo Broth

Yeast extract	1.5 g
Peptone	5 g
Lactose	12.5 g
Sodium deoxycholate	0.1 g
K_2HPO_4	4.37 g
KH_2PO_4	5 g
NaCl	5 g
Sodium lauryl sulfate	0.05 g
Na_2SO_3	2.1 g
Basic fuchsin	1.05 g
pH	7.2 ± 0.2 at 25°C
D.W.	1000 ml

Method of Preparation

1. Heat to boiling. Do not autoclave. Cool to room temperature. Dispense onto sterilised adsorbent pads.
2. Take sample volumes of 1, 4 and 15 ml and continue step 7 to 15 as described under 'procedure' earlier.
3. Incubate the plates in an inverted position at 37°C for 22-24 hrs.
4. Examine the plates, count the colonies that are pink to red with a golden metallic green sheen. Use a plate having 20-80 such colonies and not more than 200 of all types of colonies.

Faecal Coliform Test

1. Insert a sterilised absorbent pad into each of three snap lid petri plates. This has to be opened by inserting forceps into the slot between lid and base twisting the forceps.
2. Add 1-2 ml of M-Fc broth to the surface of each pad composition of M-Fc broth.

Tryptose	10 g
Proteose peptone	5 g
Yeast extract	3 g
Sodium chloride	5 g
Lactose	12.5 g
Bile salts	1.5 g
Aniline blue	0.1 g
pH	7.4

1. Add about 10 ml of 1% Rosonilic acid in 0.2 N Sodium hydroxide. Continue heating for 1 minute.
2. Cool to room temperature. Do not autoclave.
3. Use sample volumes of 1.3 to 10 ml in respective sets and begin the step No. 7 and go till step 15 as described in the 'procedure' earlier.
4. Shut sets by snapping them and seal with water proof tape, insert in a plastic bag, incubate the prepared plates in a 44.5°C water bath for 22 hrs.
5. Count blue coloured colonies with coliform characteristics. Use the plate having 20-60 colonies.

Faecal *Streptococcus* Test

1. Prepare 3 plates having KF agar.

Composition

Proteose peptone	10 g
Yeast extract	10 g
NaCl	5 g
Sodium glycerophosphate	10 g
Maltose	20 g
Lactose	1 g
Sodium azide	0.4 g
Bromocresol purple	0.015 g
pH	7.2
D.W.	1000 ml

Sterilise at 15 lbs pressure (121°C) for 10 minutes dispense in 100 ml amounts in flasks and add 1 ml of 1% unsterilised solution of TTC (Tripheny tetra sodium chloride) at room temperature.

2. Use 1, 5 and 25 ml of sample volumes and perform the steps 7 to 15 as described under 'procedure' earlier.

3. Incubate the plates at 37°C for 48 hrs.

4. Examine the plates for colonies that are light pink and flat and dark red with smooth pink margins. Count the plate that has 20-100 colonies. Calculate the faecal streptococci ratio using an average of at least 5 readings.

$$\frac{\text{No. of faecal coliform per ml}}{\text{No. of } \textit{streptococci} \text{ per ml}} = \frac{FC}{FS}$$

Interpret the result as follows :

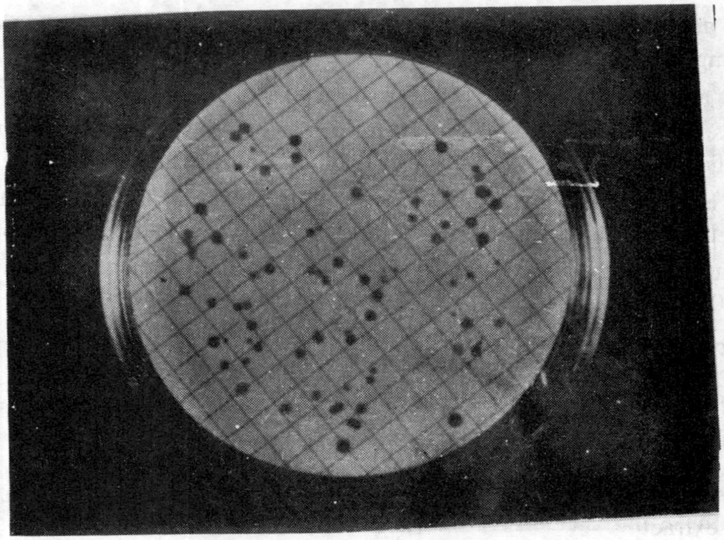

Fig. 19.3 Feacal *Streptococci* colonies.

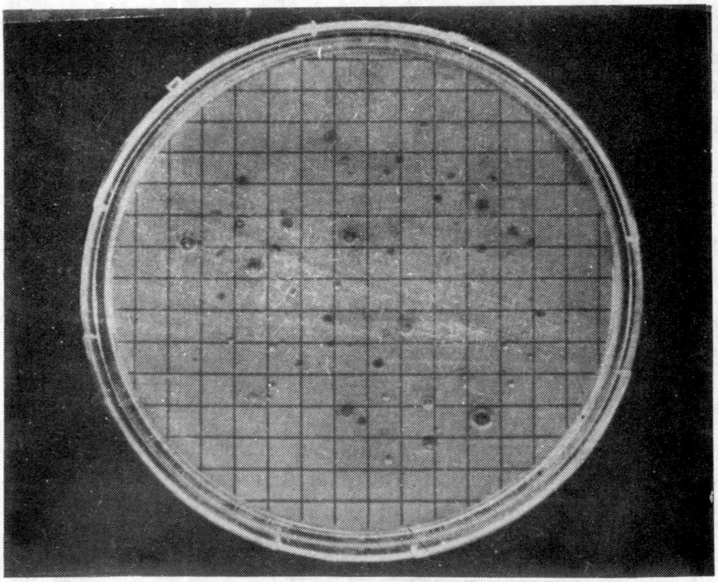

Fig. 19.4 Colonies on membrane filter.

Table 22

S. No.	FC/FS	Conclusions
1.	0.7	Pollution entirely from livestock waste.
2.	1-2	Cannot be interpreted.
3.	2-4	Predominance of human waste in mixed pollution.
4.	4	Pollution from human waste.

Air Analysis

Sampling of air is accomplished by using a membrane filter along with a sampling apparatus, or impinger. The membrane filter collects the microorganisms and supports colony growth when placed on an appropriate medium.

1. Attach the air sampling apparatus with a vacuum source.
2. Switch on the vacuum source and allow it to produce air current so that it pulls a sample of air down on to the impingement fluid. Volume of air passing through the orifice must be large enough to be measurable and representative. In general, a sample size of 10 cubic feet (280 litres) is sufficient for most indoor sampling applications.
3. Meanwhile, prepare 2 plates by placing the sterilised absorbent pad at the bottom and saturate it with 1.8-2 ml of appropriate culture medium.

Composition of yeast medium

Yeast extract	3 g
Malt extract	3 g
Peptone	5 g
Dextrose	10 g
Agar	20 g
pH	4.5
D.W.	1000 ml

4. Turn off the vacuum system after collecting the sample for 30 minutes.
5. Remove the aerosol adapter, lift the funnel corner and using a squirt bottle of sterilised water, rinse down inside walls of sterifil funnel.
6. Attach the vacuum tubing to the longer arm of aerosol adapter apply a gum rubber cap to the opening on the other side arm.
7. Switch on the vacuum so that the rubber funnel outlet cap is sucked into the receiver flask along with the impingment fluid.

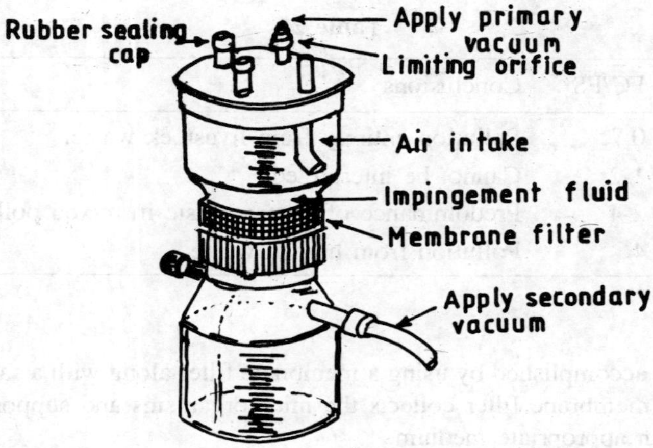

Fig. 19.5 Sterifil apparatus for sampling air.

8. Turn off the vacuum and rinse funnel walls as described above turn the vacuum back on to draw all of the rinse through the filter.
9. Switch off the vacuum. Unscrew the sterifil funnel and place it safely aside.
10. Ignite the sterilising alcohol on tips of forceps, cool and remove the filter from filter holder base.
11. Allow the extreme edge of filter to touch the extremes of pad in petridish. Now let the filter roll over the plate.
12. Seal the plate tightly, incubate at 37°C for 24 hours.
13. Observe the number of bacteria or yeast and mold colonies, perform the calculations using following formulae.

    (a) Sampling time (in minutes) × capacity of limiting orifice (litres per minute) = number of litres per sample.

    (b) $$\frac{\text{Litres of sample}}{28} = \text{Cubic feet}$$

    (c) $$\text{No. of organisms per cubic foot of air} = \frac{\text{No. of colonies observed}}{\text{Cubic feet of sample}}$$

20

Slide Agglutination Test

OUTLINE

• Precipitin test • Immunodiffusion

Principle

Rapid slide agglutination test is carried out when rapid information regarding presence of antibodies against organism is required.

Procedure

1. Take cavity slides, mark first slide as *Salmonella typhosa*. Second as *Salmonella paratyphi* A and third as *Salmonella paratyphi* B.
2. Add 2-3 drops of dense formalysed suspension of *Salmonella* antigen to the cavities.
3. To the first slide add 1 drop of unknown serum to one of the cavities and a drop of normal saline to other which will serve as a control.
4. Repeat the step 3 with second and third slide.
5. Mix the contents and heat gently for 2 minutes over a bunsen flame.

Observe agglutination and compare with control.

Precipitin Test

Principle

Here a soluble antigen reacts with its specific antibody to produce precipitate unlike agglutination where antigen are part of whole cells, and specific antibodies react with them to form a lump.

Precipitin test gives a qualitative evidence of antigen antibody reactions and limited quantitative relationships.

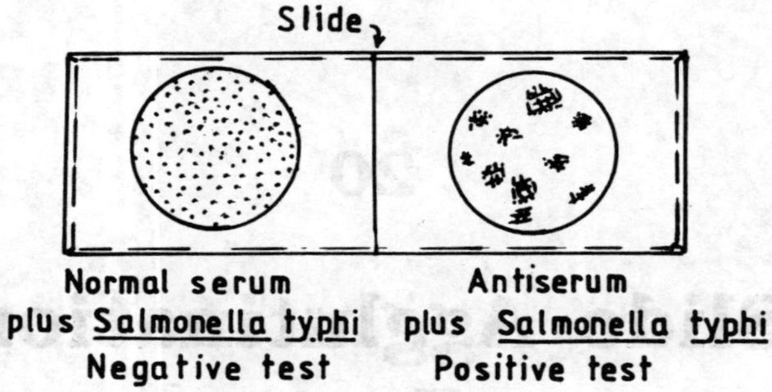

Fig. 20.1 The slide agglutination test.

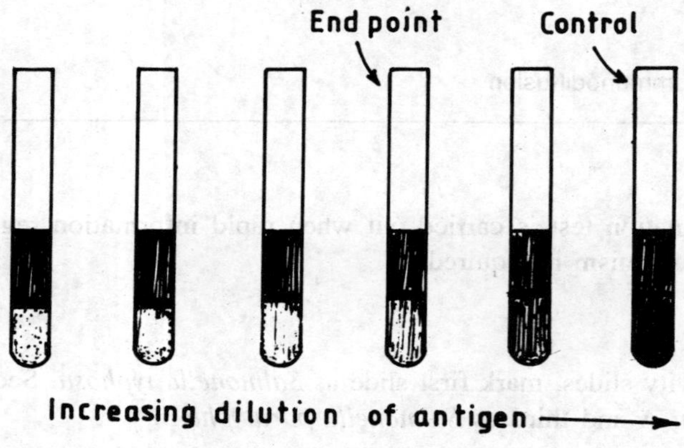

Fig. 20.2 The precipitation ring test.

Procedure

(i) Use horse serum as the antigen and rabbit antiserum for its homologus antibodies.
(ii) Prepare serial dilutions ranging from 10^{-1} to 10^{-5} of the antigen.
(iii) Transfer the dilutions to five different tubes.
(iv) Draw the antiserum in the capillary and layer it on to the antigen.
(v) Transfer a control antiserum to precipitin tube.
(vi) Observe tube for precipitin bands after 1, 5, 10, 15 and 30 minutes.
(vii) Note the highest dilution of antigen giving a precipitin band.

Immunodiffusion

Also known as Ouchterlony method. The antiserum is placed in a well cut in an agar plate the soluble antigen is introduced in a series of wells around the centre well and along the periphery of the agar plate.

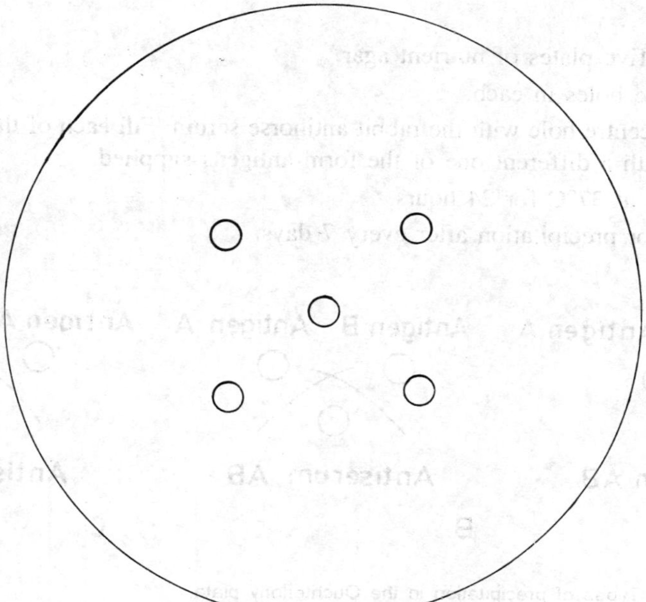

Fig. 20.3 Template for making holes in the Ouchterlony plate.

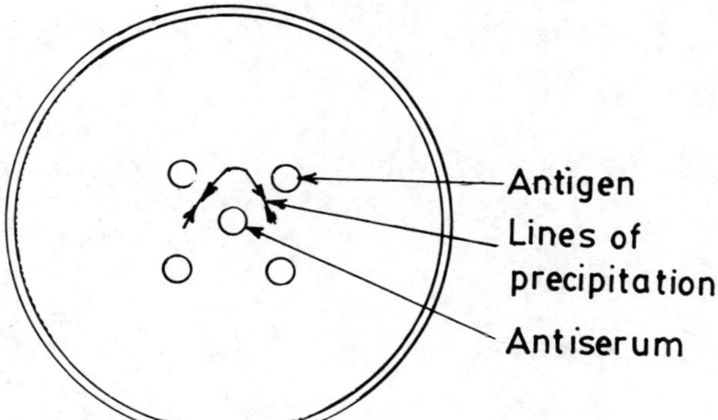

Fig. 20.4 Formation of precipitation.

Procedure

1. Prepare five plates of nutrient agar.
2. Bore five holes in each.
3. Fill the centre hole with the rabbit antihorse serum. Fill each of the 4 surrounding holes with a different one of the form antigens supplied.
4. Incubate at 37°C for 24 hours.
5. Check for precipitation after every 7 days.

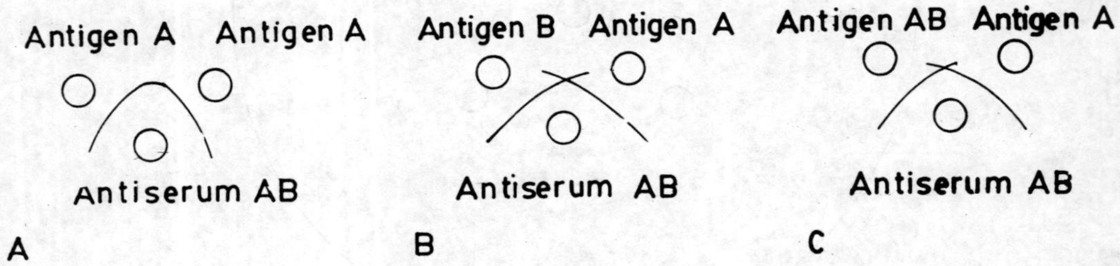

Fig. 20.5 Types of precipitation in the Ouchterlony plate.

(A) Reaction of identity, (B) Reaction of non-identity, (C) Reaction of partial identity.

21

Fungi : General Account

OUTLINE

• Definition • Mode of nutrition • Reproduction : (i) Asexual, (ii) Sexual • Sexual compatibility • Parasexuality • Classification

Definition

Fungi are thalloid achlorophyllous, usually filamentous and multicellular, with definite cell wall. They are usually non motile although the reproductive structures are motile. A prominent differentiation exists between reproductive and somatic structures.

On the basis of their mode of nutrition fungi are classified as :

1. *Parasites* – (*Gr. parasites* = eating beside another)
2. *Saprobes* – (*Gr. sapros* = rotten + bios = life)

Fungi which live on dead matter and are incapable of infecting living organisms are called *obligate saprobes.* Facultative parasites are capable of causing disease on living or dead organic matter, according to circumstances these are facultative parasites (or facultative saprobes), obligate parasites include those group of organisms which essentially require living protoplasm. Apart from saprophytic and parasitic attitude of fungus an interesting relationship exists between fungus and roots of higher plants. These structures are called as *Mycorrhiza.* (Pl. *mycorrhizae* Gr. *mykes* = mushroom + *rhiza* = root). Fungi usually store excess food in the form of glycogen or oil they prefer acidic medium for growth, usually a pH of 6. The fungal thallus consists of microscopic threads the hypha (Gr. *hyphal* = web). A hyphae is of a thin, transparent, tubular wall filled with a layer of protoplasm varying in thickness.

The protoplasm is either continuous, when there are several nuclei due to merging of septa, this condition is known as *coenocytic.* The composition of cell wall is not fixed, it may have cellulose, which is the chief constituent, in higher forms the cell wall has chitin. Callose, alignic like substance and other organic materials have also been detected in many fungi. The composition of cell wall in any fungal sp. is never same all the time.

The mass of hyphae constituting the thallus of a fungus is called mycelium. (Pl. *mycelia*, Gr. *mykes* = mushroom).

Some of the higher fungi form thick strands by their mycelium. Division of labour is exhibited by certain other fungi, here the hypha lose their identity and form *rhizomorphs* (Gr. *rhiza* = root + *morphe* = shape). Rhizomorphs are resistant to adverse conditions and remain dormant until favourable conditions return. The mycelium of parasitic fungi may be either intercellular or intracellular. An intercellular mycelium absorbs food through the host cell wall or membrane. While intracellular mycelium absorbs by means of haustoria which is probably a response to the contact stimulus as well as to stimulus of nutrients.

In saprobes the hyphae come in intimate contact with the substratum and derive the food by direct diffusion through the hyphal walls, causing a disintegration of organic matter which they utilize.

Reproduction

The formation of new individuals having all the characteristics typical of the species is accomplished in fungi by asexual and sexual reproduction. Asexual reproduction does not involve the union of nuclei, sexual reproduction, on the other hand, is characterised by the union of two nuclei.

Reproductive organs are formed either in holocarpic (where entire thallus forms reproductive organs) or eucarpic fashion (here the reproductive structures are formed from only a portion of thallus, the remainder, continues its normal somatic activities.

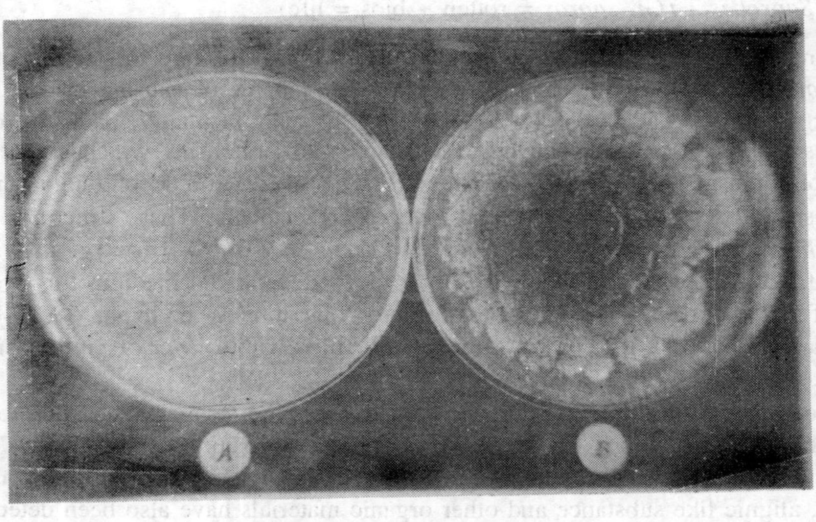

Fig. 21.1 Fungal colonies from mycelial fragmentation. (A) Colony from agar disc, (B) Days later.

Asexual Reproduction

It occurs by following means :

1. Fragmentation of hyphae into their component cells, oidia (sing. oidium) (Gr. *oidion* = small egg) or arthrospores (Gr. *arthron* = joint + *sporos* = seed, spore). If cell developes a thick wall before they separate from each other or from the parent hyphae then such spores are called *chlamydospores* (Gr. *chlamys* = mantle + *sporos* = seed). Fragmentation may also be due to external force.

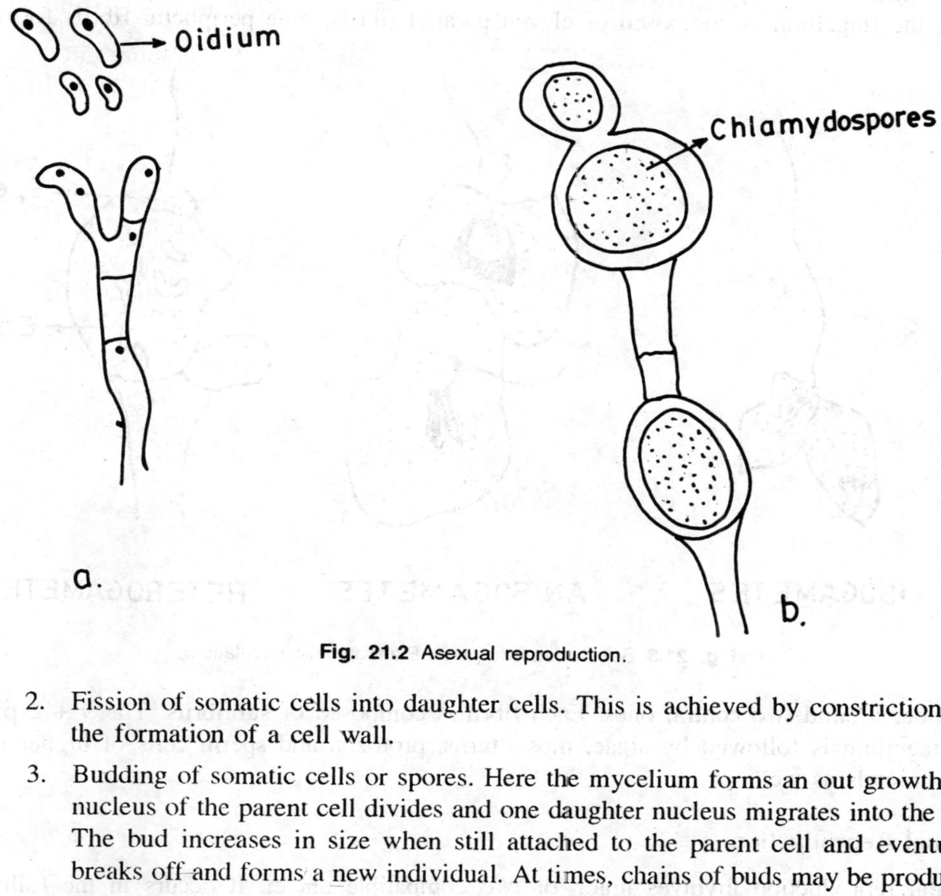

Fig. 21.2 Asexual reproduction.

2. Fission of somatic cells into daughter cells. This is achieved by constriction and the formation of a cell wall.
3. Budding of somatic cells or spores. Here the mycelium forms an out growth, the nucleus of the parent cell divides and one daughter nucleus migrates into the bud. The bud increases in size when still attached to the parent cell and eventually breaks off and forms a new individual. At times, chains of buds may be produced.
4. Production of spores : This is the most common method of asexual reproduction in fungi. Fungal spores occur in infinite variety. Fungal spores produced asexually are either borne at the tips or on the sides of hyphae such spores are referred to as *conidia* (Gr. *konis* = dust + idion). Spores produced on sporangia (Gr. *sporos* = seed, spore + angion = vessel) are called as *sporangiospores*, which are either motile or nonmotile.

Motile sporangiospores are called as *zoospores* (*zoon* = animal + *sporos* = seed, spore). If nonmotile, they are called *apalanospores* (Gr. *a* = not + *planetes* = wanderer + *sporos* = seed, spore). Zoospore bear one or two flagella, either of whiplash (lower rigid and longer part and upper short and flexible part), or tinsel type (a feathery structure consisting of long rachis with lateral hair like projections on all sides along its entire length.

The flagellar apparatus is very complex consisting of a blepharoplast (*blepharis* = eyelash + plastid) which is the base of the flagellum inside the motile cell, and a rhizoplast (Gr. *rhiza* = root + plastic) through which the flagellum is attached to the nucleus of the cell, the flagellum is composed of eleven parallel fibrils, nine peripheral fibrils forming a

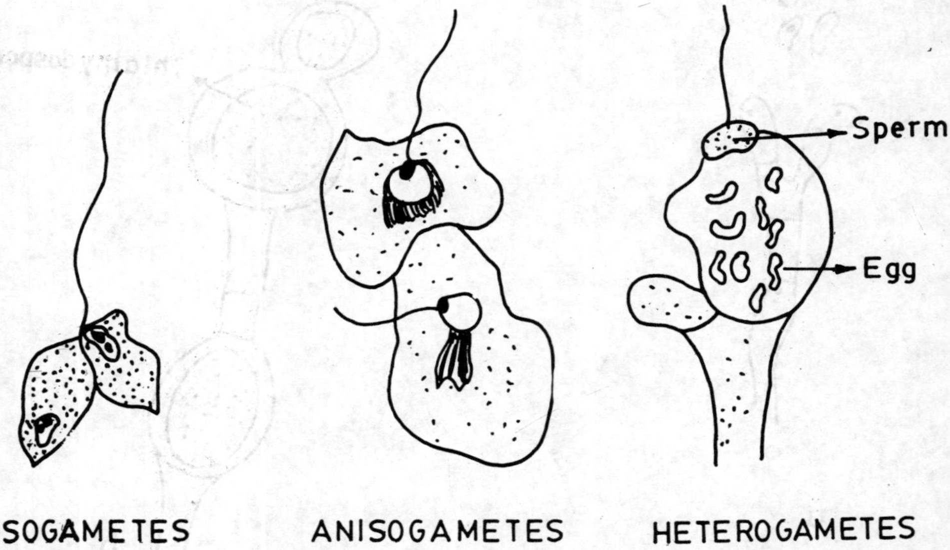

Fig. 21.3 Sexual reproduction : Planogametic copulation.

cylinder around two central ones. Each fibril is composed of subfibrils. The 9 + 2 pattern of flagellum is followed by algae, moss, ferns, protozoa and sperm cells of higher plants and animals as well.

Sexual Reproduction

Sexual reproduction involves union of two compatible nuclei. It occurs in the following phases :

1. Plasmogamy : (Gr. *plasma* = moulded object + *gamos* = marriage). It involves union of protoplasts.
2. Karyogamy : The fusion of two nuceli (Gr. *karyon* = nut, nucleus + *gamos* = marriage).

3. Meiosis : Nuclear fusion is sooner or later followed by meiosis (Gr. *meiosis* = reduction), which again reduces the number of chromosomes to the haploid.

Sex Organs

Some fungi produce distinguishable male and female sex organs on each thallus. Such species are hermaphroditic or monoecious. Some species bear distinct male and female thallus. These species are referred to as dioecious.

Sex organs are termed as *gametangium* which form sex cells – gametes. It may contain one or more gamete nuclei. Gametes may be either isogametes (morphologically indistinguishable) or heterogametes (morphologically different). In the latter, the male gametangium is called *antheridium* (Gr. *antheros* = flowery + *idion*, dinium, suffix) and the female gametangium is called *oogonium.*

Methods of Sexual Reproduction

1. Planogametic copulation : It involves fusion of two naked gametes one or both of which are motile. The motile gametes are termed as *planogametes. Anisogamous* planogametes, which are morphologically similar but different in size occur in monoblepharidales.
2. Gametangial contact : Here both male and female gametes are reduced to nucelus they are never released outside but are directly transferred to gametangium.

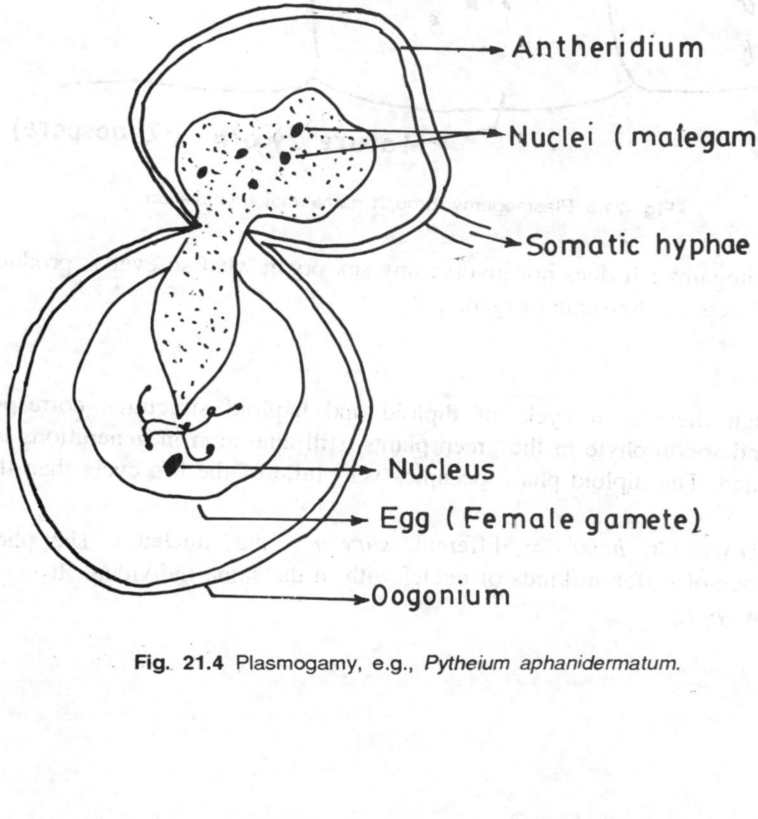

Fig. 21.4 Plasmogamy, e.g., *Pytheium aphanidermatum.*

3. Gametangial copulation : Here the fusion of entire contents of two contacting gametangia occurs in one of the two ways.
   (i) In some fungi (halocarpic ones) the entire thallus (male) pours its contents to another thallus (female) through a pore.
   (ii) Two gametangial cells fuse directly to form one, the process takes place by dissolution of the walls which are in contact of two gametangia.
4. Sepermatisation : Few fungi bear numerous, minute, uninucleate, spore like male structures the spermatia. The spermatia are carried by insects to other receptive hyphae where a pore develops to let the contents of spermatia enter the hyphae.

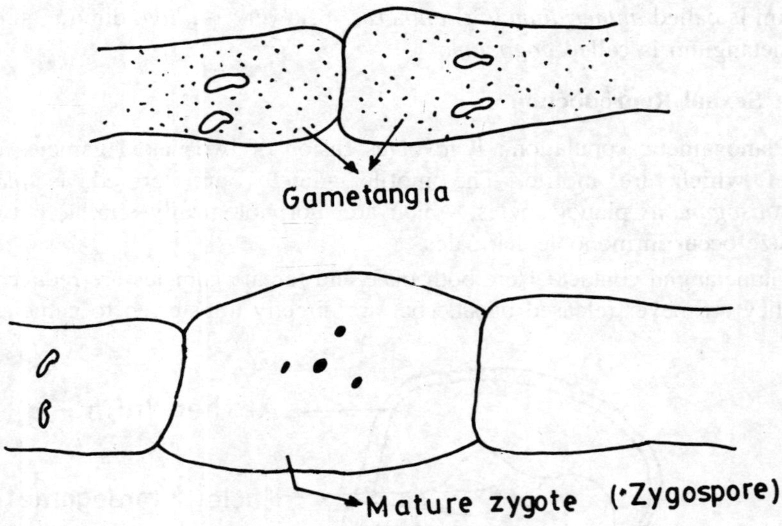

Fig. 21.5 Plasmogamy through gamatangial copulation.

5. Somatogamy : It does not involve any sex organ what so ever is produced by the fungi acts as the sexual organ.

Nuclear Cycle

In fungi, though there is a cycle of diploid and haploid structures corresponding to gametophyte and soporophyte in the green plants, still alteration in generations is not very well demonstrated. The diploid phase occupies very little of the life cycle than the haploid phase.

Heterokaryosis (Gr. *hetero* = different, *karyon* = nut, nucleus). The phenomenon, involves existence of different kinds of nuclei with in the same individual. It may originate in 4 different ways :

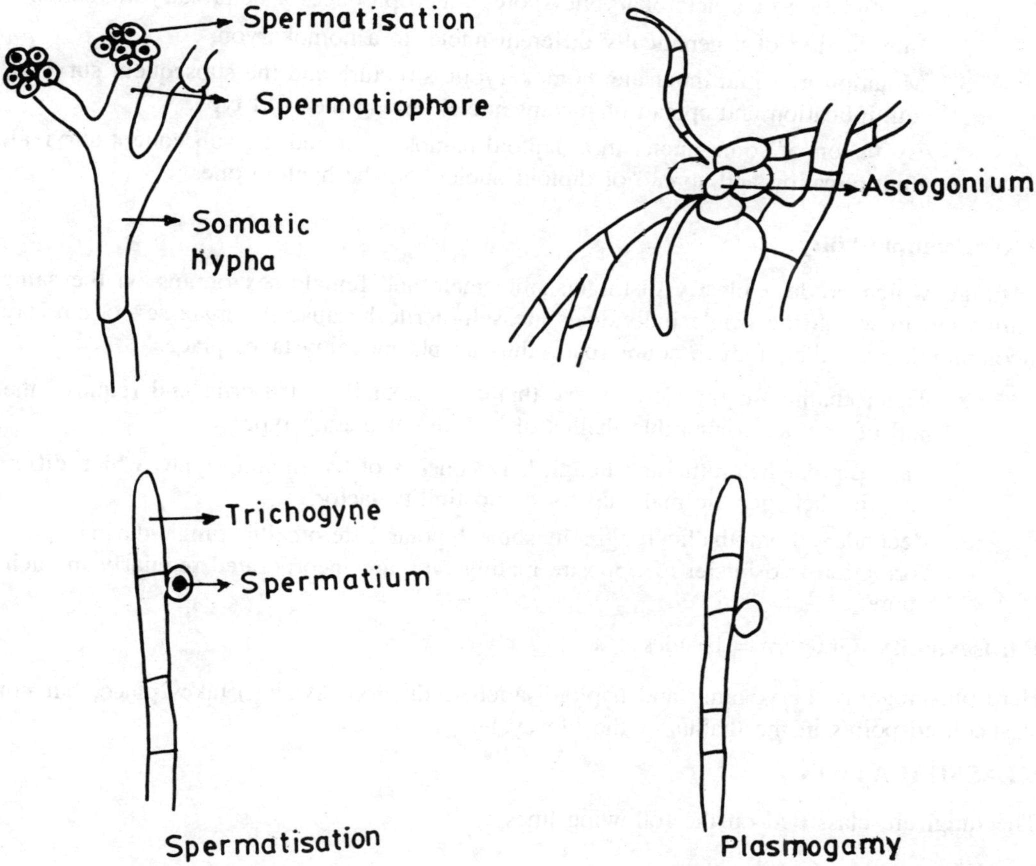

Fig. 21.6 Plasmogamy by means of spermatisation in *Podospora anserium.*

On this basis most fungi are classified into 3 categories.

A. Hermaphroditic : Each thallus bears both male and female sex organs.
B. Dioecious : Some thalli bear only male while others bear only female sex organs.
C. Sexually undifferentiated : Here, sexual functional structures are produced, which are morphologically undifferentiated as male and female sex organs.

Fungi, of the above sex categories, may be Homothallic, Heterothallic or Secondary heterothallic depending on the sexual compatibility.

1. Homothallic fungi : Here, every thallus is sexually self-fertile and therefore, reproduce sexually by itself without the aid of another thallus, therefore no dioecious fungus is homothallic.

1. Germination of a heterokaryotic spore, which produces a heterokaryotic soma.
2. Introduction of a genetically different nuclei in a homokaryon.
3. Mutation in a multinucleate, homokaryotic structure and the subsequent survival, multiplication and spread of mutant nuclei among the wild type.
4. By fusion of some nuclei in a haploid hamokaryon and the subsequent survival, multiplication and spread of diploid nuclei and the haploid one.

Sexual compatibility

In fungi, which produce clearly distinguishable male and female sex organs on the same thallus but in which the single individuals are self-sterile because the male sex organs are incompatible with their female counterparts thus no plasmogamy takes place.

2. Heterothallic fungi : Here, every thallus is sexually self-sterile and requires the aid of another compatible thallus of a different mating type.
   a. Bipolar heterothallic : Fungi, here consists of two mating types which differ in their genetic make-up for compatibility factor.
3. Secondary homothallic fungi : In some bipolar heterothallic fungi, during spore formation two nuclei of opposite mating type are incorporated regularly in each spore.

Parasexuality (Gr. *para* = besides + sex)

Here plasmogamy, karyogamy and haploidisation (reduction division) takes place, but not at specified points in the thallus or the life cycle.

CLASSIFICATION

The fungi are classified on the following lines :

Kingdom
Division
Class
Family
Genus
Species

In accordance with the recommendations of the Committee of the International Rules on Botanical Nomenclature :

Name of division of fungi ends in *mycota*, of sub-divisions in *mycotina*, of classes in *mycetes* and of sub-classes in *mycetidae*. Names of orders end in *ales*, and of families in *aceae*. Genera (Sing. genus; Latin *genus* = race) and species have no standard endings. The name of an organism is in accordance with the binomial nomenclature.

Taxanomists have divided the plant kingdom into many divisions. The fungi have been placed in the division Mycota, or Fungi. This leaves two orders Acrasials and the Labyrinthulales which represent slime molds.

Order Acrasiales : These are the cellular slime molds. The somatic structure of the Acrasiales is a simple amoeboid cell. Pseudoplasmodium (collection of amoebae) (Gr. *pseudo* = false + *plasmodium* = the soma of the Myxomycetes). Spores never bound by a common membrane. Sexual reproduction is by engulfment of one amoeba by another, and subsequent fusion of the nuclei. The fruiting bodies are sorocarps (Gr. = *seros* = heap + *karpos* = fruit).

Family Sappiniaceae

Genus *Sapinia* (1 sp.)

Family Guttulinaeceae

1. Genus *Guttulinopsis* (3 sps.)
2. Genus *Guttulina* (4 sps.)
3. Genus *Acrasis* (2 sps.)

Family Acytosteliaceae

1. Genus *Protostelium* (1 sp.)
2. Genus *Actostelium* (1 sp.)

Family Dictyosteliaceae

1. Genus *Dictyostelium* (6 sps.)
2. Genus *Polyspondylium*
3. Genus *Coenonia*

Labyrinthulales : It includes marine or fresh water parasites or saprobes. The structure is simple oval or spindle shaped. Cells secrete mucous filaments which unite to form a fine network on which the cells glide. Sexual reproduction is not known.

Genus *Labyrinthula :* It includes all marine sp. where the slime tracks are of non living material secreted by the cells, which do not feed on solid particles through phagocytosis.

Genus *Labyrinthorhiza :* It contains the fresh water species where the slime tracks are of living filaments. Some of these obtain nourishment through phagocytosis.

Division Mycota

These are the fungi. The soma varies from a unicell to an extensive mycelium. Reproduction is typically both asexual and sexual, spores being propagative units. It comprises of 2 **sub-divisions** — Myxomycotina or true slime molds, and the Eumycotina, or true fungi.

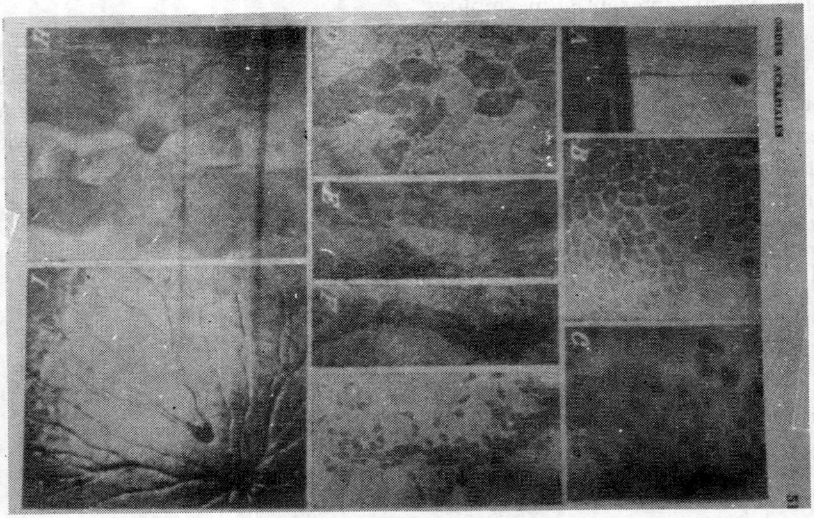

Fig. 21.7 *Dictyostelium discoidenae* : Various stages of growth.

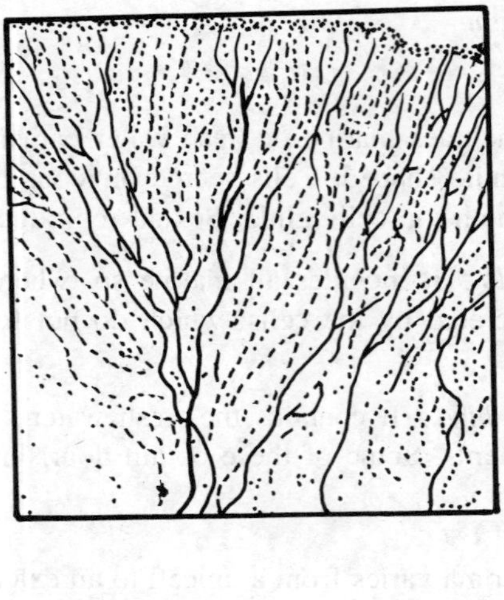

Fig. 21.8 *Labyrinthula* sp. Net plasmodium.

FLOW CHART SHOWING CLASSIFICATION OF FUNGI

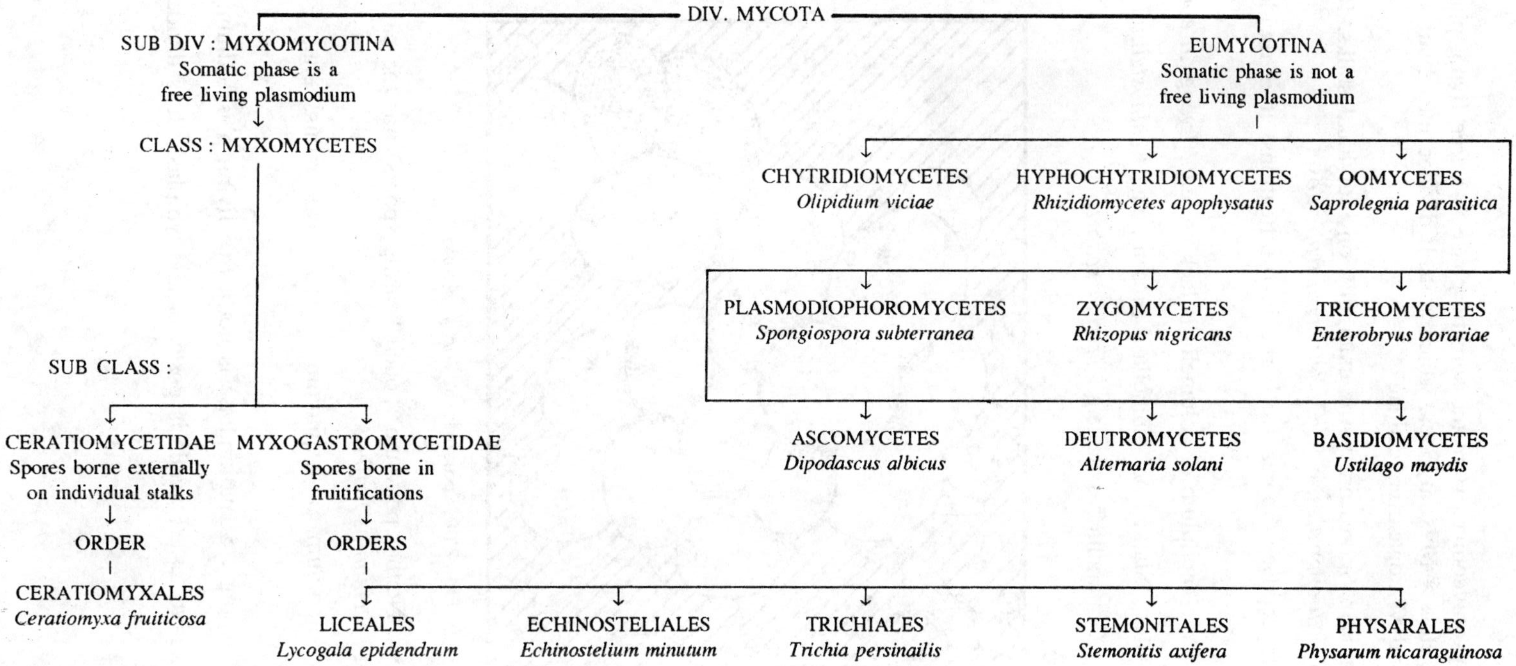

1. **Sub-division** Myxomycotina : Here the somatic phase is a free living plasmodium. It contains a single **class** – Myxomycetes, which is itself divided into 2 **sub-classes** — Ceratiomycetidae and Myxogastromycetidae.

(i) Ceratiomycetidae : Spores are borne externally on individual stalks. It contains a single **order** *Ceratiomyxales*, e.g., *Ceratiomyxa fruticulosa.*

(ii) Myxogastromycetidae : Spores are borne in fruitfications. It is divided into 5 **orders**.

1. Liceales : True capitulum or columella is lacking, e.g., *Lycogala epidendrum.*
2. Echinosteliales : Stalked or columella is present, e.g., *Echinostelium minutum.*
3. Trichiales : Columella sessile, e.g., *Trichia scabra.*

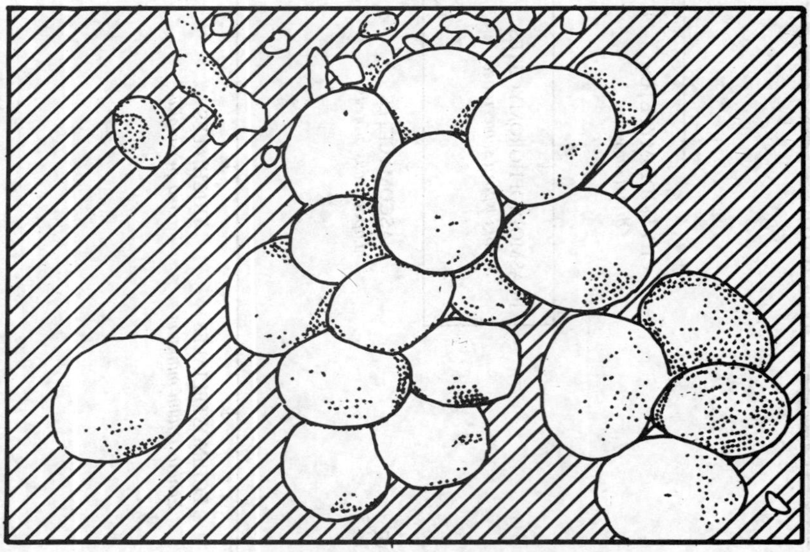

Fig. 21.9 Sporangia of *Trichia scabra.*

4. Stemonitales : Neither peridium (pouch bearing spores) nor the capitulum is calcareous, e.g., *Stemonitis axifera.*
5. Physarales : Peridium or capitulum or both calcareous, e.g., *Physarum nicarguinasa.*

2. **Sub-division** Eumycotina : Somatic phase is not a free living plasmodium.

(i) **Class** Chytridiomycetes : Flagella is posterior and whiplash. It is divided into 3 **orders.**

1. Chytridiales : True mycelium lacking; rhizomycelium present in some species. It is divided into 4 **families**.

(a) Olipidiaceae : Sexual reproduction by copulation of isoplanogametes, e.g., *Olipidium viciae.*

(b) Synchytriaceae : Reproductive organs enveloped in common membrane, e.g., *Synchytrium endobioticum.*

(c) Phyciticiaceae : Sexual reproduction by gametangial copulation, e.g., *Physophidium couchii.*

(d) Megachytriaceae : Aquatic, sporangia and resting spores, e.g., *Nowakowskiella ramosa.*

2. Blastocladiales : Sexual reproduction by fusion of planogametes, thickwalled, resistant, sporangia characteristically formed. It contains single **family** Blastocladiaceae, e.g., *Allomyces macrogynus.*

3. Monoblepharidiales : Sexual reproduction by male and female apalanogamete, no resistant sporangia. **Family** Monoblepharidiaceae, e.g., *Monoblepharis, Polymorpha.*

(ii) **Class** Hyphochytridiomycetes : Flagellum anterior, tinsel type. **Family** Rhizoidiomycetaceae, e.g., *Rhizoidiomyces apophysatus.*

(iii) **Class** Oomycetes : Reproduction asexual, zoospore biflagellate, each bearing one tinsel flagellum directed forward and one whisplash directed backward. Zoopores in sporangia of several types. It is divided into four **orders**.

1. Saprolegniales : Zoospores are borne in the sporangium. Hyphae, if present, then halocarpic/eucarpic and without constrictions. **Family** *Saprolegniaceae.*

2. Leptomitales : Zoosporia either in sporangium or in evanescent vesicle; hypae with eucarpic constrictions, e.g., *Apodachyla pyrifera.*

3. Lagenidiales : Hyphae holocarpic. **Family** Lagnenidiaceae, e.g., *Lagenidium rabenhostii.*

4. Pernosporales : Eucarpic hyphae, sporangiophore is either similar to somatic hyphae or different than it.

(a) **Family** Phythiaceae : Sporangiophore similar to somatic hyphae, e.g., *Phytophthora infestans.*

(b) **Family** Albuginaceae : Sporangia are produced in chains, e.g., *Albugo candida.*

(c) **Family** Pernosporanceae : Sporangia in clusters, e.g., *Plasmospora viticola.*

(iv) **Class** Plasmodiophoromycetes : These obligate endoparasites have their somatic phase as a plasmodium, which develops with in the host cells. Plasmodia produce zoosporangia containing zoospores produced by clevage of plasmodium. Single **family** Plasmodiophoromycetes, e.g., *Spongospora subterranea*.

(v) **Class** Zygomycetes : Sexual reproduction by gametangial copulation resulting in the formation of zygospore. Asexual reproduction by means of nonmotile spores in the form of sporangiospores or conidia. It includes 3 **orders**.

1. **Order** Mucorales : Chiefly saprobic, weakly parasitic, asexual reproduction by sporangia having one to many aplanospores, at times, conidia, e.g., *Rhizopus oryzae*. Mucorales have developed six lines of evolution. The **families** which represent each line may be grouped as follows :

(a) Mucoraceae

(b) Chonephoraceae

(c) Thamnidiaceae

(d) Pilobolaceae

(e) Mortierellaceae

(f) Syncephalastraceae

2. **Order** Entomophthorales : Parasitic, rarely saprobic, asexual reproduction by sporangia functioning as conidia forcibly discharged, e.g., *Entomophthora muscae*.

3. **Order** Zoopagales : Conidia not forcibly discharged, e.g., *Cochlonema verrucosum*.

(vi) **Class** Trichomycetes : Soma filamentous, attached to digestive tract or external cuticle of living arthropods, mycelium not immersed in host tissue. **Order** Ecrinales, e.g., *Enterobryus borariae*.

(vii) **Class** Ascomycetes : Spores resulting from **karyogamy** and meiosis, borne in asci, naked. 3 **sub-classes**.

1. Hemiascomycetidae

2. Euascomycetideae

3. Loculoascomycetidae

1. **Sub-class** Hemiascomycetidae : No ascogenous hyphae or ascocarps. It includes 2 **orders**.

1. **Order** Endomycetales : Asci arise directly from zygotes, each derived from the copulation of two cells or parthogenetically from single cell. 4 **families**.

(a) Ascosphaeriaceae : Asci in spore balls, e.g., *Ascosphaera apis*.

(b) Endomycetaceae : Asci 1-8 spored, mycelium abundant, plasmogamy by copulation of uninucleate gametangia, e.g., *Endomyces*.

(c) Spermophthoraceae : Plasmogamy by copulation of non-motile gametes released from gametangia, e.g., *Spermophthora*.

(d) Saccharomycetaceae : Mycelium scanty, e.g., *Saccharomyces cerevisiae*.

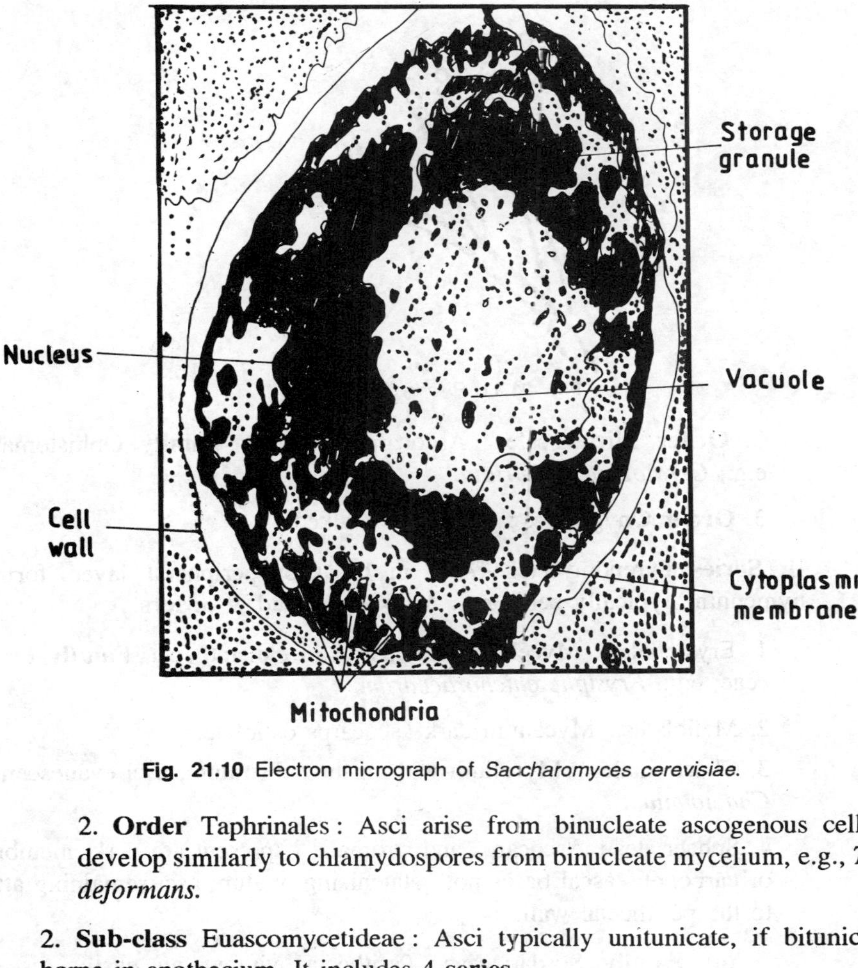

Fig. 21.10 Electron micrograph of *Saccharomyces cerevisiae*.

2. **Order** Taphrinales : Asci arise from binucleate ascogenous cells which develop similarly to chlamydospores from binucleate mycelium, e.g., *Taphrina deformans*.

2. **Sub-class** Euascomycetideae : Asci typically unitunicate, if bitunicate then borne in apothecium. It includes 4 **series**.

I. **Series** Plectomycetes : Asci scattered within the ascocarp or united in spore balls within a sporcyst. 3 **orders**.

1. **Order** Eurotiales : Ascocarps sessile and without ostiole. 3 **families.**

(a) Ascosphaeriaceae : Asci in spore balls, e.g., *Ascosphaera apis.*

(b) Gymnoascaceae : Asci not in spore balls, peridium when present, a lax network of hyphae, e.g., *Ctenomyces serratus.*

(c) Eurotiaceae : Peridium of closely interwoven hyphae usually forming a pseudoparenchyma, e.g., *Aspergillus niger* and *Penicillium notatum.*

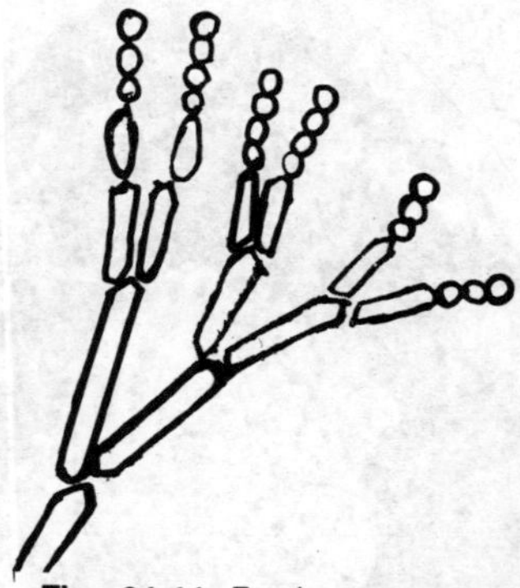

Fig. 21.11 *P. chrysogenum.*

2. **Order** Microascales : Ascocarps ostiolate. **Family** Ophistomataceae, e.g., *Ceratocystis fimbriata.*

3. **Order** Onygenales : Ascocarps stalked.

II. **Series** Pyrenomycetes : Asci in basal or peripheral layer, forming a hymenium, typically persistent. Ascocarp closed. 9 **orders.**

1. Erysiphales : Mycelium white, asci in cleistothecia. **Family** Erysiphaceae, e.g., *Erysiphe chichoracearum.*

2. Meliolales : Mycelium dark ascocarps ostiolate.

3. Chaetomiales : Mycelium with in the substratum, asci evanescent, e.g., *Chaetomium.*

4. Sphaeriales : Ascocarps and stromata, if present are dark, membranous or carbonous ascal bases not gelatenising, mature asci remaining attached to the perithecial wall.

(a) **Family** Sordariaceae : Perithecia without an easily discernible stroma, perithecia superficial, e.g., *Sordaria fimicola.*

(b) **Family** Phyllachoraceae : Perithecia immersed in the substratum, e.g., *Phyllachora graminis.*

(c) **Family** Diatrypaceae : Stroma consisting of both host and fungal tissue.

(d) **Family** Xylariaceae : Stroma entirely fungal, e.g., *Xylaria polymorpha.*

5. Diaporthales : Ascal bases gelatinising, mature asci loose in the perithecial cavity and exuding through the ostiole.

(a) **Family** Gnomoniaceae : Perithecia in substratum, e.g., *Gnomonia erythrostoma.*

(b) **Family** Diaporthaceae : Perithecium immersed in a stroma with only their beaks protruding out, e.g., *Glomerella cingulata.*

6. Hypocreales : Ascocarps and stromata if present brightly coloured, soft, fleshy or waxy. 3 **families.**

(a) **Family** Nectriaceae : Perithecia superficial, e.g., *Nectria coccinea.*

(b) **Family** Hypocreaceae : Perithecia immersed completely, e.g., *Hypomyces.*

(c) **Family** Melanosporaceae : Perithecia transparent, e.g., *Melanospora.*

7. Clavicipitales : Perithicia in stroma composed of fungal tissue. **Family** Clavicipitaceae, e.g., *Claviceps purpurea.*

8. **Order** Cronophorales : Ascocarp opening irregularly by an ostiole, but ostiole is not funnel shaped.

III. **Series** Discomycetes : Ascocarp an open apothecium as a modified form.

1. **Order** Ostropales : Ascocarps above ground and asci inoperculate.

2. **Order** Helotiales : Ascospores not thread like, sometimes needle shaped. 4 **families.**

(a) **Family** Phacidiaceae : Transitional stage between perithecia and apothecia, e.g., *Rhytisma acerinum.*

(b) **Family** Sclerotiniaceae : Apothecial initials arise from stromata and sclerotia, e.g., *Monilinia fructicola.*

(c) **Family** Geoglossaceae : Apothecia variously shaped, e.g., *Geoglossum.*

(d) **Family** Cyttariaceae : Ascocarps embedded in fleshy, globose stromata produced in large clusters, e.g., *Nothofagus.*

3. **Order** Pezizales : Operculate or sub operculate. 3 **families.**

(a) **Family** Sacroscyphaceae : Asci sub operculate, e.g., *Sacrosò-maglobosum.*

(b) **Family** Pezizaceae : Apothecia not differentiated into stipe and pileus, e.g., *Peziza vesiculosa.*

(c) **Family** Halvellaceae : Apothecia differentiated clearly into stipe and pileus, e.g., *Helvella esculenta.*

4. **Order** Tuberales : Ascocarps below the ground, e.g., *Tuber aestivum.*

IV. **Series** Laboulobeniomycetes : Specialised exoparasites of insects or arachinids, thallus usually limited. **Order** Laboulobeniales, e.g., *Laboulbenia formicarum.*

3. **Sub-class** Loculoascomycetidae : Asci bituniate, ascocarp and ascostroma present. It is divided into 5 **orders.**

1. **Order** Myriangiales : Locules uniascal, usually scattered at various levels in ascostroma. 2 **families.**

(a) **Family** Elsinocaceae : Four celled hyaline ascospores mature in globose asci, e.g., *Elsinoe veneta.*

(b) **Family** Myriangiaceae : The tissues surrounding the cavities in the centre of the ascocarp is of ascogenous hyphae, e.g., *Piedraia hortai.*

2. **Order** Dothideales : Absence of interascal threads in locule and formation of bitunicate asci. 3 **families.**

(a) **Family** Dothideaceae : Spermatia borne in spermogonia, e.g., *Mycosphaerella fragariae.*

(b) **Family** Pseudosphaeriaceae : Ascostroma is a pseudothecium in which asci are formed single with in the ascocarp and remain separated by stromatal tissue, e.g., *Pseudoplea gaumanii.*

(c) **Family** Capxodiaceae : Individual ascostroma resemble perithecia.

3. **Order** Pleosporales : Stroma not boat shaped. Pseudoparaphyses present. 3 **families.**

(a) **Family** Venturiaceae : Plant parasites, two celled ovoid/ellipsoid ascospores, e.g., *Venturia inequalis.*

(b) **Family** Pleosporaceae : Plant parasite, e.g., *Pleospora.*

(c) **Family** Lophiostomataceae : Ascocarp bears the general appearance of a perithercium with a prominent neck.

4. **Order** Microthyriales : Locules polyascal arranged in a basal layer, stroma flattened and dimidiate, e.g., *Microthyrium microscopicum*.

5. **Order** Mysteriales : Boat shaped or branched ascocarps, with a long slit like opening along the entire length of fruiting body, e.g., *Glonium*.

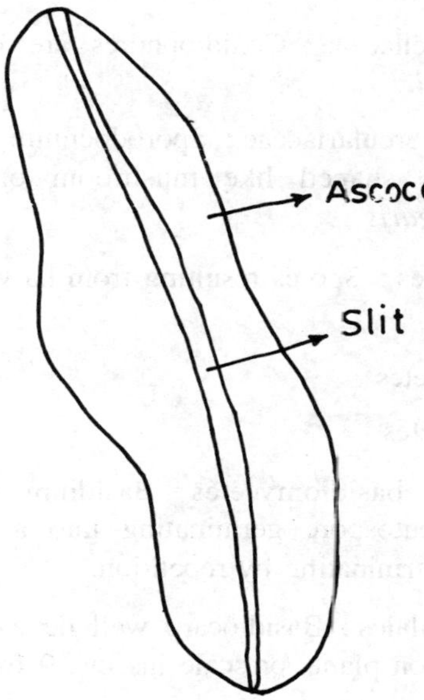

Fig. 21.12 Ascocarp of *Glonium* sp

(viii) **Class** Deuteromycetes : These fungi lack a sexual phage (perfect stage) hence are popularly known as fungi Imperfecti. These may be considered as conidial stages of Ascomycetes whose sexual stages do not exist. 3 **orders**.

1. **Order** Sphaeropsidales : Reproduction by means of conidia borne in pycnidia. 2 **families**.

(a) Sphaeropsidaceae : Small, ostiolar pycnidia sunk in the substratum very short conidiophores and hyaline spherical or oval spores, e.g., *Phyliosticta acericola*.

(b) Zythiaceae : Pycnidia are light or brightly coloured and their walls are soft or waxy, e.g., *Zythia fragariae*.

2. **Order** Melanconiales : Conidia when produced, borne in acervuli, e.g., *Gloeosporium musarum*.

3. **Order** Moniliales : Reproduction by means of conidia borne otherwise, by oidia or by budding.

(a) **Family** Cryptococceae : Yeast like fungi.

(b) **Family** Moniliaceae : Unorganised, conidia on hyaline conidiophores or hyaline hyphae, e.g., *Trichophyton megninii.*

(c) **Family** Dematiaceae : Hyphae and conidia are typically dark, e.g., *Alternaria solani.*

(d) **Family** Stibellaceae : Conidiophores are united into synnemats, e.g., *Ceratiocystis ulmi.*

(e) **Family** Tuberculariaceae : Sporodochium or stromata covered with conidiophores is shaped like mushroom or is crescent shaped, e.g., *Tubercularia vulgaris.*

(ix) **Class** Basidiomycetes : Spores resulting from karyogamy and meiosis borne on basidia. 2 **sub-classes.**

1. Heterobasidiomycetes
2. Homobasidiomycetes

1. **Sub-class** Heterobasidiomycetes : Basidium septate or deeply divided, consisting of a teleutospore germinating into a promycelium, basidiospores usually capable of germinating by repetition.

1. **Order** Tremellales : Basidiocarp well developed, mostly saprobic, some species parasitic on plants or scale insects. 9 **families.**

(a) Ceratiobasidiaceae

(b) Tulasnellaceae

(c) Sirobasidiaceae

(d) Dacrymycetaeae

(e) Tremellaceae

(f) Hyaloriaceae

(g) Phleogenaceae

(h) Auriculariaceae

(i) Septobasidiaceae

2. **Order** Uredinales : Teleutospores present, plant parasites, basidiospores discharged forcibly. 3 **families.**

(a) **Family** Puccinaceae : Teleutospores free or variously united, but never in layers or crusts, e.g., *Puccnia graminis.*

(b) **Family** Melanosporaceae : Teleutospores laterally united into layers, crusts or columns *Cronatium rificola.*

(c) **Family** Coleosporiaceae : Teleutospores become septate during germination without the formation of an external promycelium, e.g., *Coleosporium solidaginis.*

3. **Order** Ustillaginales (Smuts) : Basidispores sessile, not forcibly discharged. 3 **families.**

(a) **Family** Ustillaginaceae : Promycelium septate; basidiospores produced laterally from each cell of the promycelium, e.g., *Ustilago maydis.*

(b) **Family** Tilletiaceae : Promycelium non-septate, basidiospores on a terminal cluster, e.g., *Tilletia caries.*

(c) **Family** Graphiolaceae

2. **Sub-class** Homobasidiomycotes : Basidium not septate or deeply divided; basidiospores usually germinating by germ tube.

1. **Order** Exobasidiales : Basidiocarp lacking, hymenium covering the surface of the parasitized plant tissues, e.g., *Exobasidium vaccinii.*

I. **Series** Hymenomycetes : Hymenium present and exposed before the spores are mature. 2 **orders.**

1. **Order** Polyporales : Hymenium borne in various ways, but if basidia line pores or gills, texture of basidiocarp not soft and putrescenet. 6 **families.**

(a) Thelephoraceae : Basidiocarp cobwebby membranous and hard, e.g., *Streum frustulatum.*

(b) Clavariaceae : Basidiocarp club shaped. Usually erect, e.g., *Clavariadelphus pistillasis.*

(c) Cantharellaceae : Basidiocarp mushroom like or funnel shaped, e.g., *Cantharellus cibarius.*

(d) Hydnaceae : Hymenium covering pendant warts, spines or teeth, e.g., *Hericium coralloides.*

(e) Meruliaceae : Hymenium lining interior of pits or tubes, pits shallow, fertile on ridges.

(f) Polyporaceae : Tubes deep, if shallow, sterile on ridges, texture not soft and putrescent, e.g., *Polyporus sulphureus.*

2. **Order** Agaricales : Hymenium borne on gills, or if lining the interior of pores, then basidiocarp soft and putrescent. 5 **families.**

(a) Boletaceae : Hymenophore made up of tubes, e.g., *Boletus edulis*.

(b) Paxillaceae : Hymenophore of lamellae which are often united by veins.

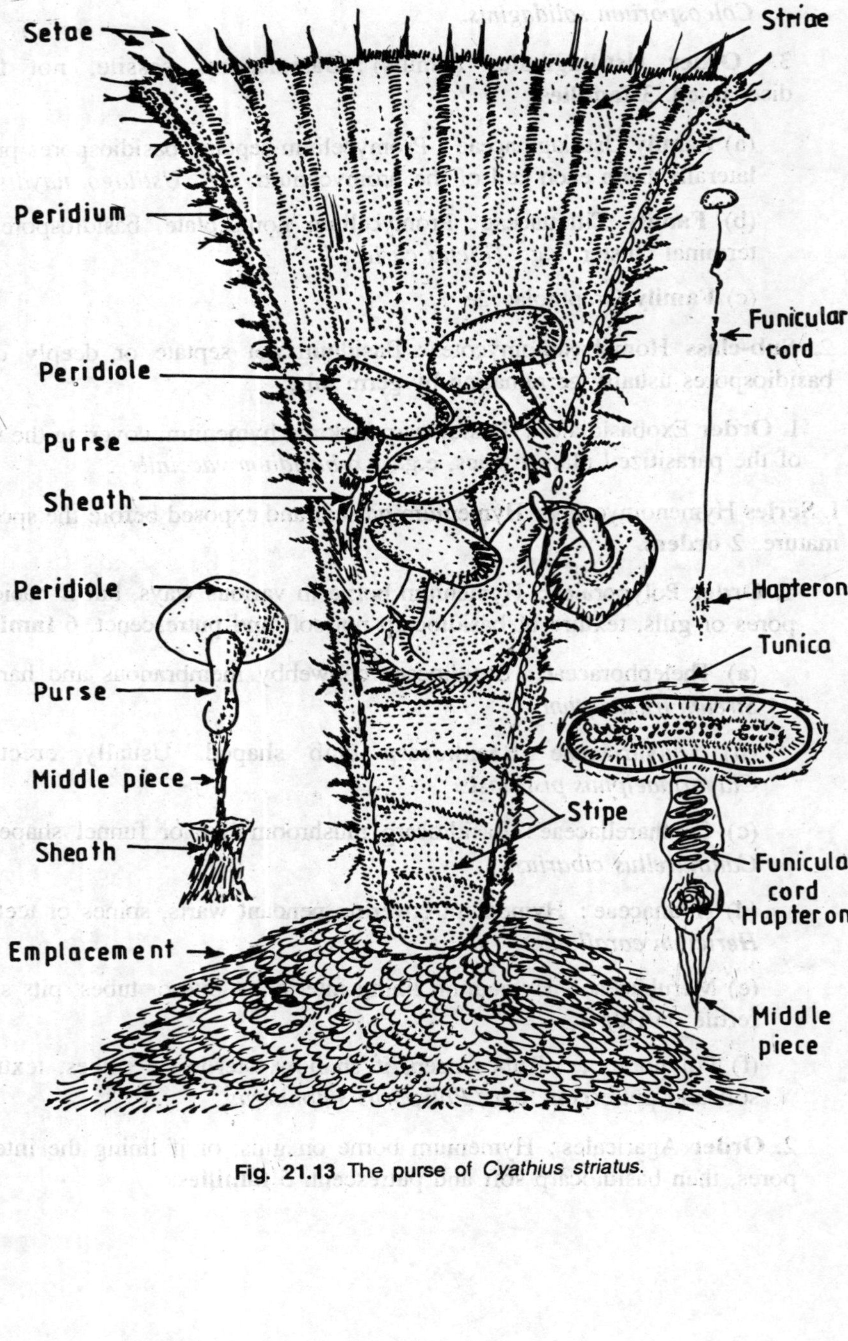

Fig. 21.13 The purse of *Cyathius striatus*.

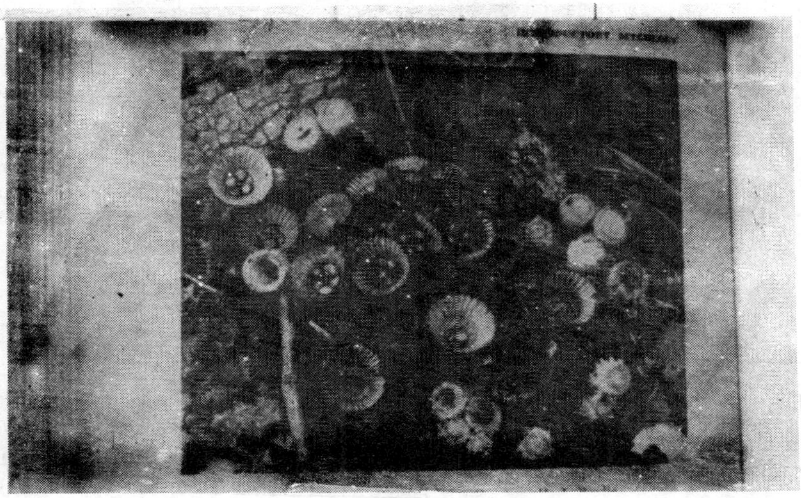

(c) Russulaceae : Basidiocarp fleshy content of sporophore containing nests of sphaerocysts, e.g., *Russula emetica.*

(d) Hygrophoraceae : Gills, waxy, broadly triangular in section.

(e) Agaricaceae : Gills not waxy, narrow in section *Amanita verna.*

II. **Series** Gasteromycetes : Basidiocarps remain closed at least till the spores have been released from the basidia. 5 **orders.**

1. **Order** Hymenogastrales : Gleba fleshy to cartillagenous, or if slimy then not exposed to maturity, e.g., *Endoptychum agaricoides.*

2. **Order** Lycoperdales : Hymenium present in early stages, spores lightly coloured and small, e.g., *Clavatia gigantea.*

3. **Order** Sclerodermatales : Hymenium lacking or indistinct, spores mostly dark large, e.g., *Astraeus hygrometricus.*

4. **Order** Phallales : Gleba exposed on a receptacle, e.g., *Dictyophora duplicata.*

5. **Order** Nidulariales : Glebal chambers forming waxy peridioles, or entire gleba separating as a unit from peridium, e.g., *Cyathus striatus.*

22

Lab Exercises on Fungi

OUTLINE

• Isolation and identification of fungi • Study of cellular slime molds • Composition of some synthetic and non-synthetic media • Yeast : morphology and reproduction • Micrometry

I. Isolation and Identification of Fungi

1. Prepare 6 plates of potato dextrose agar. (See overleaf for the composition of media.)
2. Label them as A, B, C, D, E & F.
3. Inoculate A & B with soil, C & D with water and expose the remaining ones to air.
4. Each sample should be diluted serially as described in Chapter (3)
5. Incubate at 37°C for 48 hrs.
6. Examine the plate cultures for gross morphology.

II. Cellular slime molds

As described earlier, cellular slime molds are not true fungi. A typical example of this group is *Dictyostelium discoidium*. It starts its life cycle with the dispersal of spores from a sporangium. Upon the absorption of water, the wall of the spore cracks open and a single amoeba like cell, a *myxamoeba,* emerges. These cells live independently, as they move about feeding on bacteria, they grow and reproduce by fission. This, free living, feeding stage ends following depletion of nutrients, the myxamoeba stream to an aggregation centre and forms a *Pseudoplasmodium.* The pseudoplasmodium migrates as a unit, leaving behind a trail of slime. When the migration ceases, the cells of the pseudoplasmodium differentiate to form a sporangium with spores. The cells from the front, or apical, region of the slug organise to form a stalk. Cells from posterior region, move up the stalk as a ball of cells and become tip of the stalk.. These cells transform into spores, each capable of initiating another life cycle of this slime mold.

Procedure

1. Prepare a nutrient agar plate. Using a wax pencil draw two lines forming a right angle on the bottom of the agar plate.
2. Streak a loopful of *E. coli* on the agar surface along each of the lines marked on the bottom of the plate.
3. Inoculate a loopful of *D. discoidium* at one end of each bacterial streak. Incubate the plate at room temperature for 2 days.
4. After incubation, examine the growth on the plate with the naked eye and with a hard lens. It is advisable to observe the plate on the stage of a microscope and examine under the 4 and 16 mm objectives. If an aggregation is visible disrupt the aggregate with a needle to observe the component cells. Repeat the procedure with a pseudoplasmodium. If no fruiting bodies are detected, reincubate the culture for 2 days, and then re-examine.
5. *Dictyostelium* can be isolated from the sample using Dictyostelium medium. The composition is given below :

Dictyostelium medium

Proteose peptone	10 g
Yeast extract	5 g
Glucose	10 g
Sodium acetate*	0.35 g
KH_2PO_4	0.35 g
pH	6.4-6.6

* In place of sodium acetate 2.3 ml of glacial acetic acid can be used.

COMPOSITION OF SOME COMMON LABORATORY MEDIA

I. NON SYNTHETIC MEDIA

1. Corn Meal Agar

Corn meal	20 g
Peptone (if desired)	20 g
Dextrose (if desired)	20 g
Agar	15 g
D.W.	1000 ml

2. Oat Meal Agar

Oats	100 g
Agar	15 g

Fig. 22.1 Migration of cellular slime mold slugs on an agar surface.

D.W. 1000 ml

3. Potato Dextrose Agar

Peeled potato 250 g
Glucose 20 g
Agar 15 g
D.W. 1000 ml
pH 6-6.5

4. Soil Extract Agar

Soil extract 100 ml
Glucose 1 g
K_2HPO_4 0.5 g
Agar 15 g
Tap water 900 ml
pH 7-7.2

Preparation of Soil Extract

Steam 1000 g garden soil in one litre of tap water in an autoclave at 15 lbs. pressure for 30 mins. Add a small quantity of calcium carbonate and filter the soil suspension through double layers of filter papers. The filterate is sterilised in a flask and stocked, for future use.

5. Malt Extract Agar

Useful for the growth of wood destroying and many other fungi.

Malt extract	25 g
Agar	15 g
D.W.	1000 ml

6. Martin's Rose bengal Streptomycin Agar

Dextrose	10 g
Peptone	5 g
KH_2PO_4	1 g
$MgSO_4$	0.5 g
Rose bengal (1 part in 30,000 parts of medium)	
Agar	20 g
Streptomycin	0.03 g
D.W.	1000 ml

7. Waksman's medium

Specially used for counting soil fungi

Glucose	10 g
Peptone	5 g
K_2HPO_4	1 g
$MgSO_4$	0.5 g
Agar	20 g
D.W.	1000 ml
pH	4

8. Sabraud's Medium

Medium is suitable for many fungi including human or animal pathogens.

Glucose (or Maltose)	40 g
Peptone	10 g

Agar	15 g
D.W.	1000 ml

II. SYNTHETIC MEDIA

1. Czapek's solution (for soil inhabiting fungi)

Sucrose	30 g
Sodium nitrate	2 g
K_2HPO_4	1 g
$MgSO_4$	0.5 g
KCl	0.5 g
$FeSO_4$	0.01 g
D.W.	1000 ml

2. Coon's medium (For *Fusarium*)

Saccharose	7.2 g
Dextrose	3.6 g
$MgSO_4$	1.23 g
K_2HPO_4	2.72 g
KNO_3	2.02 g
D.W.	1000 ml

Add 25 ppm of malachite green or 40 ppm gention Violet.

3. Richards solution (for *Aspergillus*)

KNO_3	10 g
KH_2PO_4	5 g
$MgSO_4$	2.5 g
$FeCl_3$	0.02 g
Sucrose	50 g
D.W.	1000 ml
pH	6.6-7.2

YEAST – MORPHOLOGY AND REPRODUCTION

Morphology : Vegetative cell

1. Prepare wet amounts by suspending a loopful of yeast culture in a water iodine

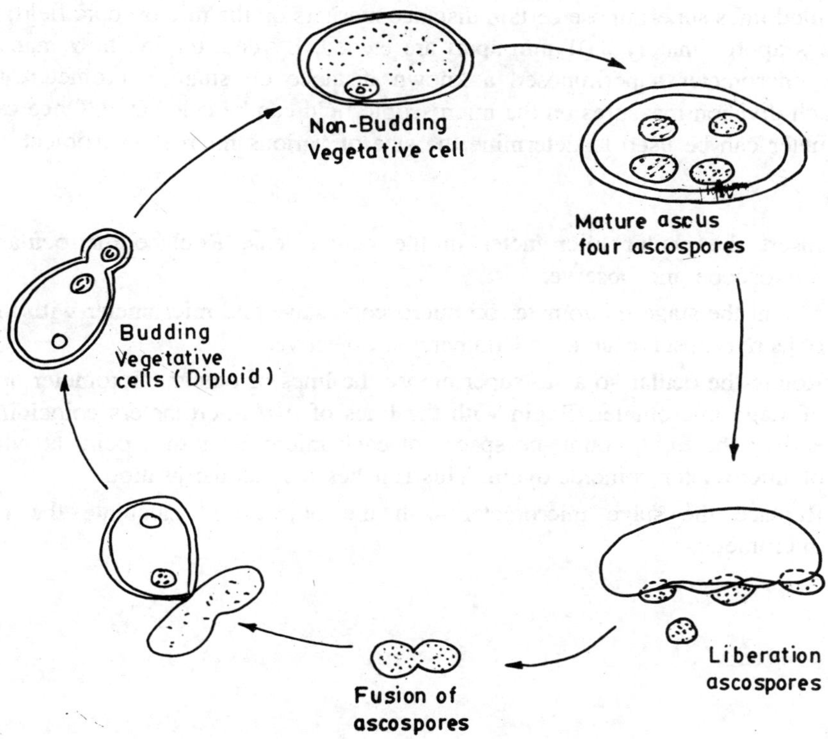

Fig. 22.2 Life cycle of yeast.

solution (3 drops water to one drop of gram's iodine) and layer each with a coverslip.

2. Observe each preparation under high-objective, noting the morphology of the yeasts, including any visible internal structures. Scan several microscopic fields to find budding cells.

Sexual Reproduction : Ascospore

1. From the yeast grown on glucose acetate medium provided, prepare stains of the culture by the malachite green spore staining technique, but do not use heat.
2. Observe under the oil-immersion objective for ascospores.

Micrometry

Measurement of minute objects seen under the microscope is called Micrometry. The unit of such measurement is called micron (μ or micrometer). One micrometer equals to 1/1000 of a mm. The ocular or eye-piece micrometer consists of a diaphragm with graduated scale unit. The ruled lines superimpose certain distance markers on the microscopic field, on which parallel lines approximately 0.01 mm apart are etched. By determining how many units of the ocular micrometer superimposed a known distance on stage micrometer, the exact distance each division measures on the microscopic field can be calculated. Once caliberated the micrometer can be used to determine the size of various microscopic objects.

Procedure

1. Insert the ocular micrometer in the ocular lens. Replace the ocular lens in microscope and observe.
2. Mount the stage micrometer on microscope stage and micrometer with low power objective observe under oil immersion objective.
3. Rotate the ocular so as to superimpose the lines of ocular micrometer upon those of stage micrometer. Begin with the lines of two micrometers coinciding at one end of the field, count the spaces of each micrometer to a point at which lines of micrometer coincide again. This finishes the standardisation.
4. Replace the stage micrometer with the object and calculate the results in micrometers.

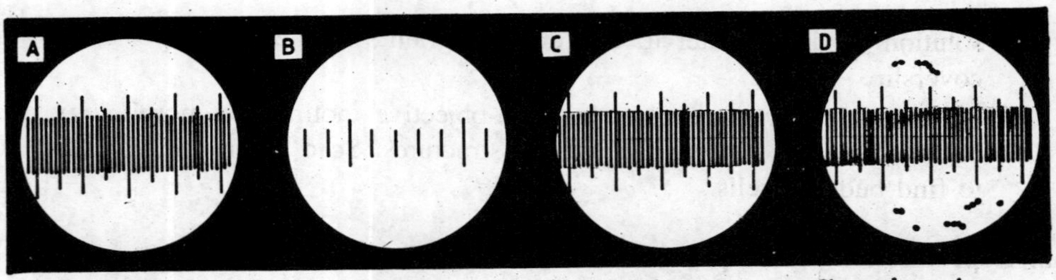

A Ocular micrometer in microscope eyepiece; divisions arbitrary.

B Stage micrometer distances absolute; each division equals 10 micrometers.

C The ocular micrometer superimposed on the stage micrometer. In this lens system, what is the absolute length in micrometers covered by 10 ocular divisions ?

D Stage micrometer (scale B) is now replaced by a slide. From your calculations in C, what is the length of the chain of streptococci ?

Fig. 22.3 Standardization of the ocular micrometer and its use in measuring bacterial cells (1 micron = 1 micrometer [µm] = 1 × 10^6 metres).

23

Cyanobacteria

OUTLINE

• Introduction • Some media compositions for cyanobacteria

Introduction

Cyanobacteria are oxygenic phototrophic bacteria. They exhibit an enormous diversity of shapes and arrangements, from unicellular cocci or rods to long trichomes. Cyanobacteria are wide spread in soil, fresh water and marine habitats while some are thermophilic. Some cyanobacteria grow in symbiosis with other organisms. For example, they may occur as

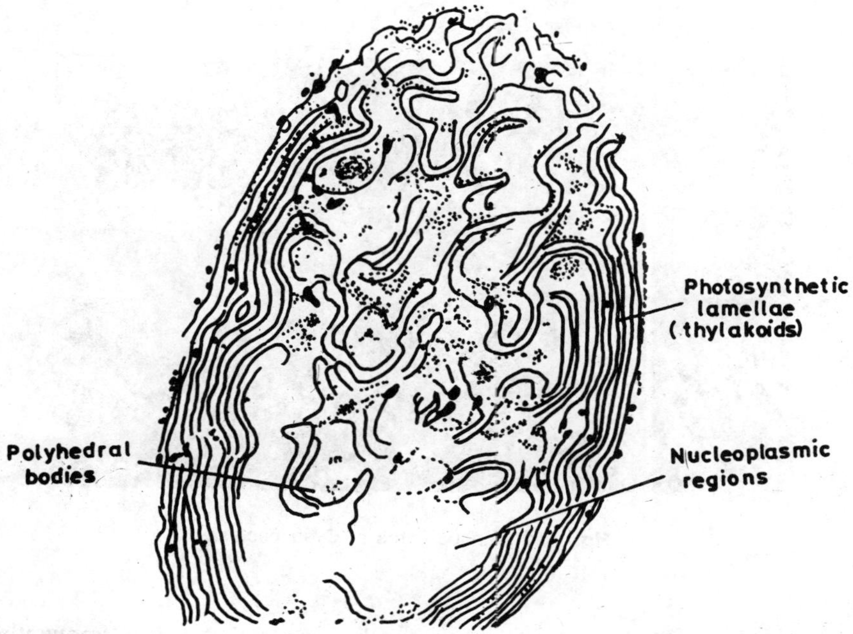

Fig. 23.1 Section of a vegetative cell of the cyanobacterium Anabaena azollae. Most of the photosynthetic lamellae (thylakoids) are peripheral, but some extend into the mid-portions of the cell.

algal symbionts or lichens. Some live within the plant bodies of liverworts, waterferns, cyacdes where they fix nitrogen. Cyanobacteria have also been found to be associated with certain protozoa, where they are called cyanellae.

Cyanobacteria contain chlorophyll rather than bacteriochlorophyll. Because of this cells absorb red light of 680 to 683 nm. Other pigments include water insoluble carotenoids and also water soluble phycobilins, which are the major light-absorbing pigments in cyanobacteria that can transmit the energy of absorbed light to the chlorophyll. Blue phycobilins occur in all cyanobacteria and absorb light at wavelengths between 500 and 650 nm. A red phycobilin, phycoerythrin, occurs in few species and absorbs shorter wavelengths between 470-600 nm. Cyanobacteria possessing phycoerythrin have a red or brown color instead of usual bluish green.

Cyanobacteria use H_2O as an electron donor for CO_2 fixation, some bacteria can also use H_2S as an electron donor in a manner similar to that used by green sulphur bacteria.

The photosynthetic apparatus (i.e., chlorophyll, carotenoid pigments, photochemical reaction centres and photosynthetic electron chain) is contained in thylakoids – flattened membrane sacs located within the cell. Phycobilisomes bearing phycobilin pigments are studded on these thylacoids.

Some trichome forming cyanobacteria can fix N_2, e.g., *Anabaena.*

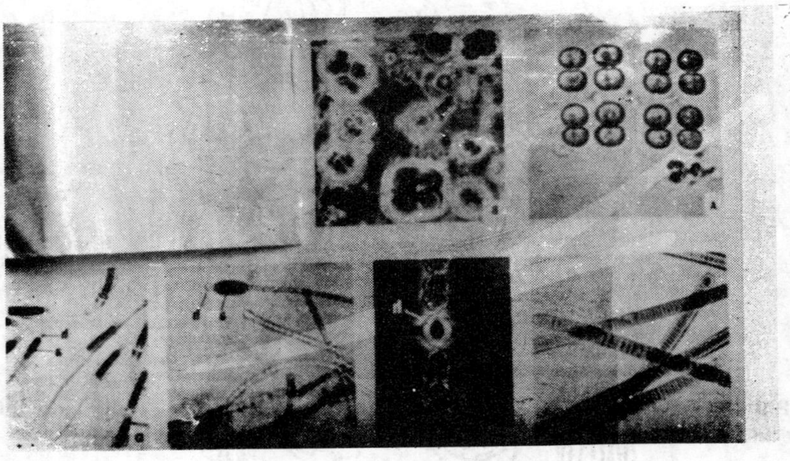

Fig. 23.2 Various forms of Cyanobacteria

Cyanobacteria or the blue green algae as they are called, occur frequently in lakes, ponds, boggs etc. Samples collected from such places should be transferred to the freshly prepared medium for the desired cyanobacterium.

After growth, cultures should be maintained viable at room temperature in non direct light. Following table shows various methods of preserving blue green algae.

Organisms	Methods used
1. *Tolypothrix tenius* and *Calothrix brevisimia*	Lyophilised with serum albumin.
2. *Nostoc* sp.	(a) Dried with sand.
	(b) Freezing at –20°C.
	(c) Freezing at various temperatures.
	(d) Lyophilized.
	(i) Air dried on special earth.
	(ii) Lyophilised with various additions.
	(e) Mineral oil

SOME MEDIA COMPOSITIONS FOR CYANOBACTERIA

Allen (1968)
Modified from Hughes et al (1958)

Composition	g/l
$NaNO_3$	15
K_2HPO_4	0.03
$MgSO_4 . 7H_2O$	0.075
Na_2CO_3	0.02
$CaCl_2$	0.027
$Na_2SiO_3 . 9H_2O$	0.058
E.D.T.A.	0.001
Citric acid	0.006
Ferrous citrate	0.006

Microelements 1 ml per litre of a solution having g/l

H_3BO_3	2.86
$MnCl_2 . 4H_2O$	1.81
$ZnSO_4 . 7H_2O$	0.222

$Na_2MoO_4 . 2H_2O$	0.391
$CuSO_4 . 5H_2O$	0.079
$Co (NO_3)_2 . 6H_2O$	0.0494
pH	7.8

Employed for *Anacystis nidulans; Gleocapsa alpicola.* Extremely valuable for surface culture and assists the growth of single celled colony when 1.5% agar is used.

Cannon et al (1971)

Modified agar Safferman and Morris (1964)

Composition	g/l
$NaNO_3$	0.17
$Ca (NO_3)_2 . 4H_2O$	0.06
K_2HPO_4	0.01
$MgSO_4 . 7H_2O$	0.025
Na_2CO_3	0.02
$Na_2SiO_3 . 5H_2O$	0.044
Ferric citrate	0.35
Citric acid	0.35
$CuCl_2 . 2H_2O$	0.001
$ZnCl_2$	0.04
H_3BO_3	0.24
$CoCl_2 . 6H_2O$	0.02
$MnCl_2 . 4H_2O$	0.014

Employed for *Plectonema boryanum.*

Castenholz (1970)

Composition	g/l
Nitrilotriacetic acid	0.1
$CaSO_4 . 2H_2O$	0.06
$MgSO_4 . 7H_2O$	0.1
NaCl	0.008
KNO_3	0.103

$NaNO_3$	0.689
Na_2HPO_4	0.111
$FeCl_3$ solution	0.2905 g (1 ml)*
Micro element solution	(0.5 ml)

Composition	g/l
$MnSO_4 . H_2O$	2.28
$ZnSO_4 . 7H_2O$	0.50
H_3BO_3	0.50
$CuSO_4 . 2H_2O$	0.025
$Na_2MoO_4 . 2H_2O$	0.045
Conc. H_2SO_4	0.50

* Values in parentheses indicate the amount of solution in 1000 ml of media.

pH before autoclaving 8.2 and 7.5-7.6 after sterilisation. Suitable for thermophilic fungi (45°C and 70°C).

Gerloff et al. (1950)
Modified after Chu (1945)

Composition	g/l
$Ca (NO_3)_2 . 4H_2O$	0.04
K_2HPO_4	0.01
$MgSO_4 . 7H_2O$	0.025
Na_2CO_3	0.02
$Na_2SiO_3 . 9H_2O$	0.025
Ferric citrate	0.003
Citric acid	0.003

Suitable for various species of blue green algae including *Nostoc muscorum*.

Gorham et al (1964)

ASM	μ moles/125 ml
$NaNO_3$	1000
$MgSO_4$	200

$MgCl_2$	200
K_2HPO_4	100
$FeCl_3$	2
H_3BO_3	10
$MnCl_2$	7
$ZnCl_2$	0.8
$CoCl_2$	0.02
$CaCl_2$	0.0002
Na_2 EDTA	20

ASM-1	µ moles/125 ml
$NaNO_3$	2000
$MgSO_4$	200
$CaCl_2$	200
K_2HPO_4	100
Na_2HPO_4	100
$FeCl_3$	4
H_3BO_3	40
$MnCl_2$	7
$ZnCl_2$	3.2
$CoCl_2$	0.08
$CuCl_2$	0.0008
Na_2 EDTA	20

Employed for toxic strains of *Anabaena* including *A. flos-aquae.*

Van Baalen (1965)
Dm, modified after Kratz and Myers (1955)

Composition	g/l
$MgSO_4 . 7H_2O$	0.25
K_2PO_4	1.00
$Ca (NO_3)_2 . 4H_2O$	0.025
$NaNO_3$	1.00
Na_2 EDTA	0.031
$Fe (SO_4)_2 . 6H_2O$	0.004

NH_4Cl	0.1
Microelement solution	0.6 ml
Composition	g/l
$ZnSO_4 . 7H_2O$	8.82
$MnCl_2 . 4H_2O$	1.44
MoO_3	0.71
$CuSO_4 . 5H_2O$	1.57
$Co (NO_3)_2 . 6H_2O$	0.49

Employed for unicellular coccoid and unicellular blue green algae especially for the quantitative surface plating of *A. nidulans* and hence accurate estimation of viable counts.

Kratz and Myers (1955)

Medium 'C'	g/l
$MgSO_4 . 7H_2O$	0.25
K_2HPO_4	1
$Ca (NO_3)_2 . 4H_2O$	0.025
KNO_3	1
Sodium citrate $2H_2O$	0.004
Micro elements	1 ml
Composition	g/l
H_3BO_3	2.86
$MnCl_2 . 2H_2O$	1.81
$ZnSO_4 . 7H_2O$	0.222
MoO_3 (85%)	0.0177
$CuSO_4 . 5H_2O$	0.079

Medium 'D'	g/l
$MgSO_4 . 7H_2O$	0.15
K_2HPO_4	0.01
$NaNO_3$	1
EDTA	0.05
$Fe_2 (SO_4)_3 . 6H_2O$	0.004

NH_4Cl	0.1
Microelement solution	0.6 ml
Composition	g/l
$ZnSO_4 . 7H_2O$	8.82
$MnCl_2 . 4H_2O$	1.44
MoO_3	0.71
$CuSO_4$	1.57
$Co (NO_3)_2 . 6H_2O$	0.49

Employed for unicellular coccoid and unicellular blue green algae especially for the quantitative surface plating of *A. nidulans* and hence accurate estimation of viable counts.

Medium 'Z' after Zenader	g/l
$NaNO_3$	0.467
$Ca (NO_3)_2 . 4H_2O$	0.059
K_2HPO_4	0.031
$MgSO_4 . 7H_2O$	0.025
Na_2CO_3	0.021
Micro elements	10 ml
Composition	g/l
EDTA	0.08
$FeSO_4 . 7H_2O$	0.01

Allen and Arnon Medium (A & A)

Growth media	mls of stock solution/litre	
	–Pi	+Pi
A & A	25	6.25
A & A/4	6.25	3.1
A & A/8	3.1	1.56

A & A undiluted is used for plates. The dilutions are used for liquid growth. A & A/4 is used for all *Nostoc*, A & A/8 for *Anabaena*.

For cultures of combined nitrogen, add a final concentration of 2.5 mM NH_4Cl or 5 mM $NaNO_3/KNO_3$ in an equimolar mix. All combined nitrogen cultures are buffered with 5 mM MOPS. Nitrogen and buffer are added from sterilised stocks first before inoculation.

Notes :

1. Shake the –Pi stock before each use.
2. Add –Pi and +Pi stocks to double distilled water. If they are mixed without water insoluble precipitates will form.
3. Check pH before autoclaving; it should be 7.8. If it is less, the +Pi stock should be remade.
4. For plates or slants, add 1-1.3% purified agar.
5. Note that the dilutions (A & A/4, A & A/8) have increased phosphate.

A & A stock solution

Make the following stock solutions and store them at 4°C. To assemble the –Pi stock, combine them 1 : 1 : 1 : 1 (and store it at 4°C).

1. $MgSO_4$ stock
   $MgSO_4 . 7H_2O$ — 20 g/500 ml
2. $CaCl_2$ stock
   $CaCl_2 . 2H_2O$ — 6 g/500 ml
3. NaCl — 20 g/500 ml
4. Micro element stock

Add in order listed. Wait for each to dissolve. Always shake before use.

1.	Double distilled water	1090 ml
2.	A & A . Fe . EDTA solution (See below)	160 ml
3.	$MnCl_2 . 4H_2O$	360 ml
4.	MoO_3 (Molybdic acid) 85% purity or	36 mg
	$Na_2MoO_4 . 2H_2O$ (99% purity)	66.1 mg
5.	$ZnSO_4 . 7H_2O$	44 mg
6.	$CuSO_4 . 5H_2O$	15.8 mg
7.	H_3BO_3 (boric acid)	572 mg
8.	NH_3VO_3 (NH_4 + Metavanadate)	4.6 mg
9.	$CoCl_2 . 6H_2O$	8 mg

A & A Fe-EDTA solution

1. Dissolve 5.2 g KOH pellets in 186 ml double distilled water. Add 20.4 g Na_2 EDTA. $2H_2O$.

2. Dissolve 13.7 g $FeSO_4 . 7 H_2O$ in 364 ml double distilled water (or use $Fe_2 (SO_4)_3 . nH_2O$. Ferric sulphate, *n* hydrate)
3. Mix solutions 1 and 2.
4. Bubble millipore filtered air (0.45 mm) through the combined solution until the colour changes (darkness to brownish orange may take 4 min or 4 hr). This makes 550 ml Fe EDTA solution.
5. Store at 4°C.

+Pi stock solution

$K_2HPO_4 . 3H_2O$ – 28 g/500 ml

24

Viruses

OUTLINE

• Isolation of bacrteriophage • Bacteriophage production • Virus culture in embryonating chicken eggs • Isolation of tobacco mosaic virus • Cultivation of viruses in tissue culture • Viral haemagglutination

Viruses are ultramicroscopic particulate macromolecules composed primarily of a nucleic acid genome, either DNA or RNA and protein. These obligate parasites, whose nucleic acid or genome controls and utilizes the synthetic capacity of their host cells for replication occur morphologically either as rodshaped virions, cubical or polyhedral virions and those possessing more complex structures.

Basically, a viral particle, or virion, is a nucleic acid core, surrounded by a protein coat, or capsid, composed of protein subunits or capsomeres. In some more complex virus the nucleocapsid is surrounded by an additional envelope and some have spike like surface appendages.

Viruses which infect bacteria are referred to as bacteriophages and can be isolated from natural environments. Animal and plant viruses can be isolated from infected tissues.

1. Isolation of Bacrteriophage

1. Add 45 ml of raw sewage to 5 ml of nutrient broth.
2. Inoculate the mixture with 5 ml of a 24 hour culture of *E. coli* and incubate at 37°C for 24 hrs.
3. Centrifuge at 2500 rpm for 10 minutes.
4. Allow the supernatant to pass through a bacteriological filter.
5. Inoculate 0.1 ml of 24 hour culture *E. coli* used in step 2 to each of the 3 tubes having 3 ml of soft agar (0.7%).
6. Prepare 3 nutrient agar plates and label them as 1, 2, 3. To one of the tubes of inoculated soft agar add one drop of sewage culture filterate. Mix and pour on a nutrient agar plate no. 1.
7. Add 5 drops of sewage culture filterate to the second tube and pour the mixture on the plate no. 2.

8. Pour the third tube of soft agar on plate No. 3. This will serve as a control.
9. Allow the soft agar to solidify and incubate the plates (inverted at 37°C until plaques are visible (6 to 24 hrs).
10. Isolate a pure culture of phage by cutting out a single plaque with a sterilised loop and transferring it to a young culture of the original host *E. coli.*
11. Incubate the infected *E. coli* and an uninfected control at 37°C with gentle shaking for 3 hrs. Observe for clearing of the infected tube.

2. Bacteriophage Production (Single step growth)

1. Plate out an *E. coli* strain B culture at dilutions of 10^{-5}, 10^{-6}, and 10^{-7}.
2. Pipette 0.9 ml of *E. coli* into a sterilised test tube.
3. Pipette out 0.1 ml of a T_2 bacteriophage suspension in 0.9 ml of *E. coli* wait for 5 minutes, so that the phage particles are absorbed by 80% of the bacterial cells.
4. Pipette the phage and *E. coli* mixture in a sterile centrifuge tube and centrifuge at 5000 rpm for 20 mins. Decant the supernatant, resuspend the pellet of cells in 0.9 of nutrient broth and mix well by swirling. This removes unabsorbed phage particles.
5. Add 0.1 ml of the above cell suspension to 9.9 ml of nutrient broth and mix well by swirling.
6. Inoculate 0.1 ml of the above diluted cell suspension to 9.9 ml of nutrient broth. Mix well by swirling. Incubate at 37°C in the water bath provided.
7. Melt five tubes of regular nutrient agar and five tubes of soft nutrient agar (0.7% agar) temper the agar to 45-50°C and pour a tube of regular nutrient agar into each of five petriplates, and allow to harden.
8. Temper the soft agar tubes to 45°C, add 3-4 drops of original *E. coli* strain B culture to each tube.
9. Consider the step 3 to be zerotime. Add 0.1 ml sample from diluted culture of step 6 to a tube of the inoculated soft agar at intervals of 20, 25, 30, 35 and 40 minutes. Mix each tube and pour on the surface of one of the nutrient agar plates prepared in step 7. Allow the soft agar to solidify and incubate the plates (inverted) at 37°C.
10. Determine the count of *E. coli* from the plates made in step 1. Examine the plates from step 8 and count and record number of plaques at the various time intervals.

3. Virus Culture in Embryonating Chicken Eggs

Viruses are obligate parasites, studies of the culture of animal viruses showed the developing chick embryos serve as a cultivating medium for many of them. Avian-embryo culture is economical and convenient than use of laboratory animals, this is commonly used for the isolation, identification, titerating and maintenance of many animal viruses as well as for the production of vaccines.

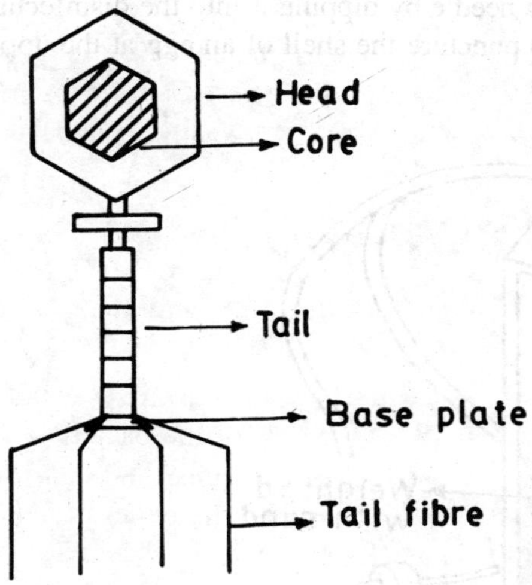

Fig. 24.1 (a) Bacteriophage.

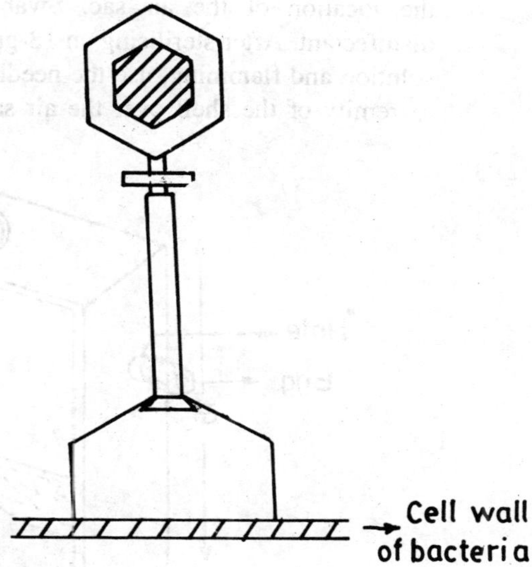

Fig. 24.1 (b) Phage fibres attack to cell wall of host.

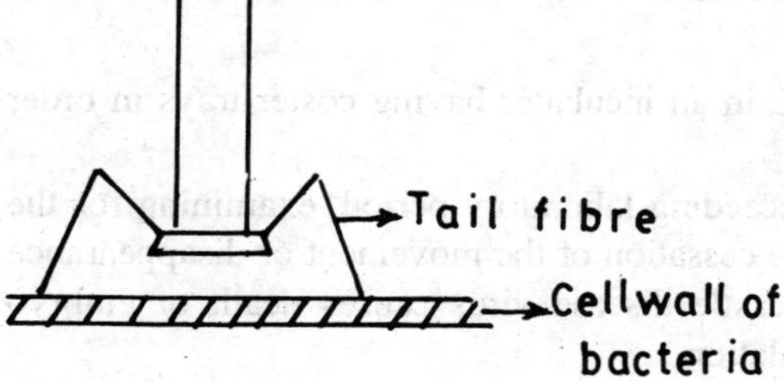

Fig. 24.1 (c) Tail contacts cell wall.

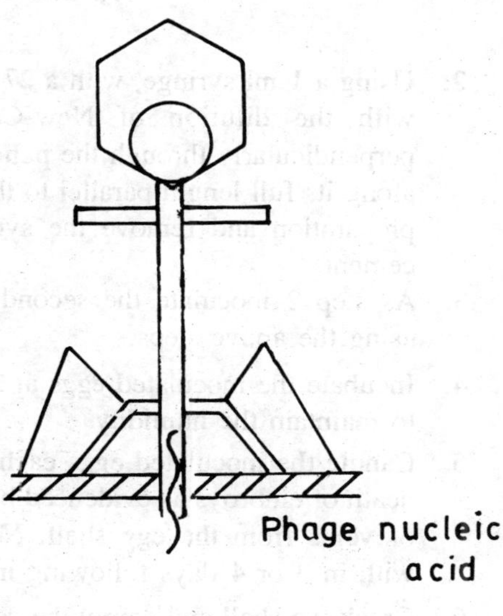

Fig. 24.1 (d) Tail sheath contracts and DNA is injected through cell wall to the exterior surface of the membrane

1. Candle two embryonated eggs with the long axis in the horizontal plane. Mark the location of the air sac. Swab the marked area with the iodine-alcohol disinfectant. After sterilising an 18-gauge needle by dipping it into the disinfectant solution and flamming, use the needle to puncture the shell of an egg at the upper extremity of the shell over the air sac.

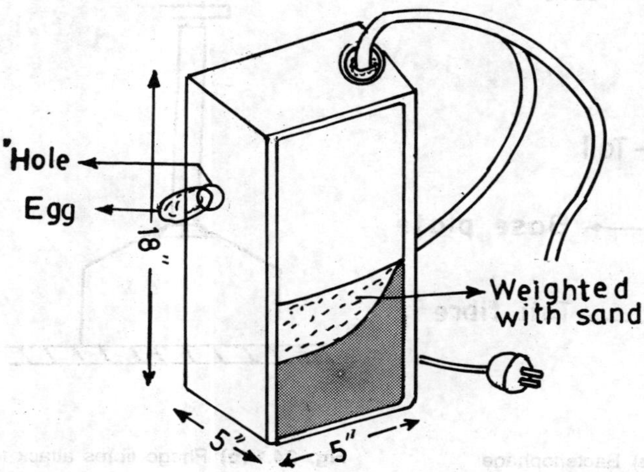

Fig. 24.2 Candling an egg.

2. Using a 1 ml syringe, with a 27 gauge, 3/4" needle, inoculate the allantoic cavity with the dilution of New-Castle disease virus by inserting the needle perpendicularly through the puncture in the shell, passing the needle into the egg along its full length parallel to the long axis of the egg. Inject 0.2 ml of the virus preparation and remove the syringe needle. Seal the puncture hole with duco cement.
3. As step 2 inoculate the second embryonated egg with 0.2 ml of sterile saline using the above steps.
4. Incubate the inoculated eggs at 37°C in an incubater having coster trays in order to maintain the humidity.
5. Candle the inoculated eggs each succeeding laboratory period, examining for the death of embroys as evidenced by the cessation of the movement or disappearance of veins from the egg shell. New castle disease virus causes death of embryo with in 3 or 4 days following inoculation.
6. Crack the shell and empty the contents into a petriplate when the embryo is dead. Compare the appearance of the two embryo.

Note : New-Castle disease virus can cause conjunctivitis in humans and hence it is advisable to wash hands carefully with soap before and after experiment.

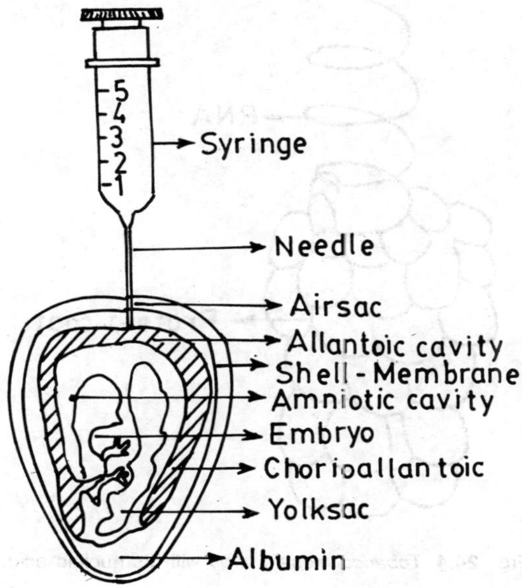

Fig. 24.3 Inoculation of virus into the allantoic cavity.

4. Isolation of Tobacco Mosaic Virus

1. Remove the leaves from diseased tobacco plant. Grind the leaves in a sterilised waring blender having 100-150 ml of sterilised water.
2. Filter the particles by pouring the contents of the blender through a filter paper.
3. Pass the filterate through a bacteriological filter.
4. Keep a healthy tobacco plant beside. Select the leaves which are to be inoculated. Wash them with carborundum using a wash bottle.
5. Rub the leaf with a piece of cotton that has been soaked in the filterate from step 3.
6. Wash the carborundum and excess virus from the leaf with distilled water from wash bottle.
7. Keep aside, a control set.
8. Check the plants during each laboratory period upto 3 weeks. Look for both localised and systemic symptoms.

Tobaccomosaic virus can also be isolated from some commercial products. Russian and Turkish brand of cigarettes are good sources of these viruses.

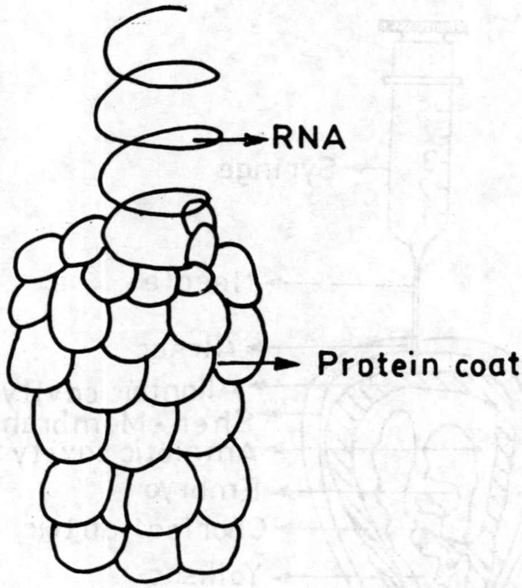

Fig. 24.4 *Tobacco mosaic virus* with its nucleic acid.

5. Cultivation of Viruses in Tissue Culture

Tissue culture is the in vitro propogation of cells. A convenient source for tissue is the **chick** embryo. Following the infection and proliferation of cells virus causes them to retract, to become round and to lyse. Some viruses cause the formation of plaques in tissue culture.

Procedure

Study the morphology of chicken fibroblast monolayers the preparation is described in following steps.

1. Place the chicken embryo in phosphate buffered saline (P.B.S.) composition :

A.	NaCl	8 g	B.	Na_2HPO_4	1.15 g
	KCl	0.2 g		KH_2PO_4	0.2 g
	$CaCl_2 . 2H_2O$	0.13 g		D.W.	200 ml
	$MgCl_2 . 6H_2O$	0.1 g			
	D.W.	800 ml			

Autoclave separately and mix when cool.

2. Cut tissue into pieces of about 1 mm in size.
3. Wash fragmented sections several times in fresh phosphate buffered saline.

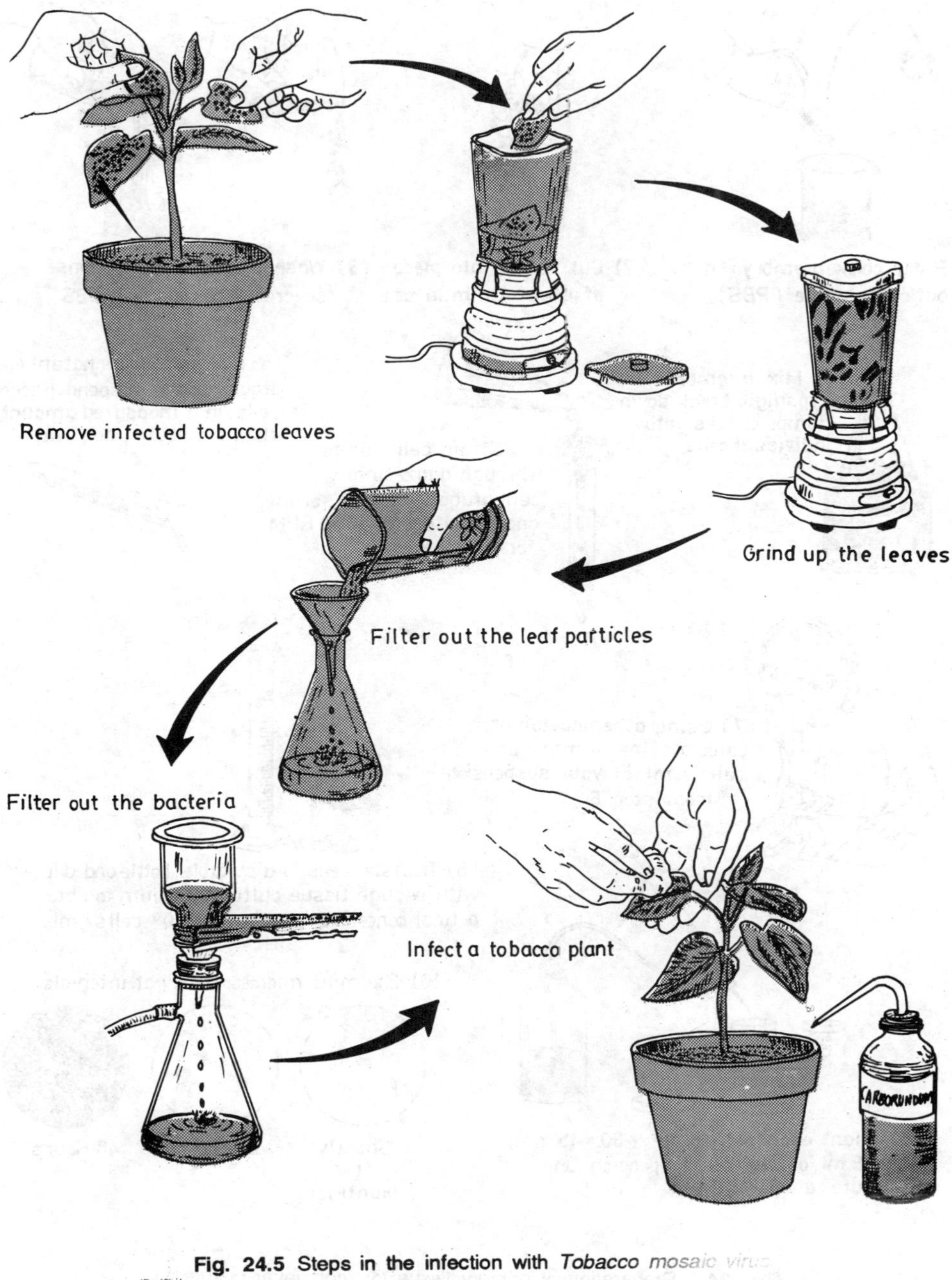

Fig. 24.5 Steps in the infection with *Tobacco mosaic virus*

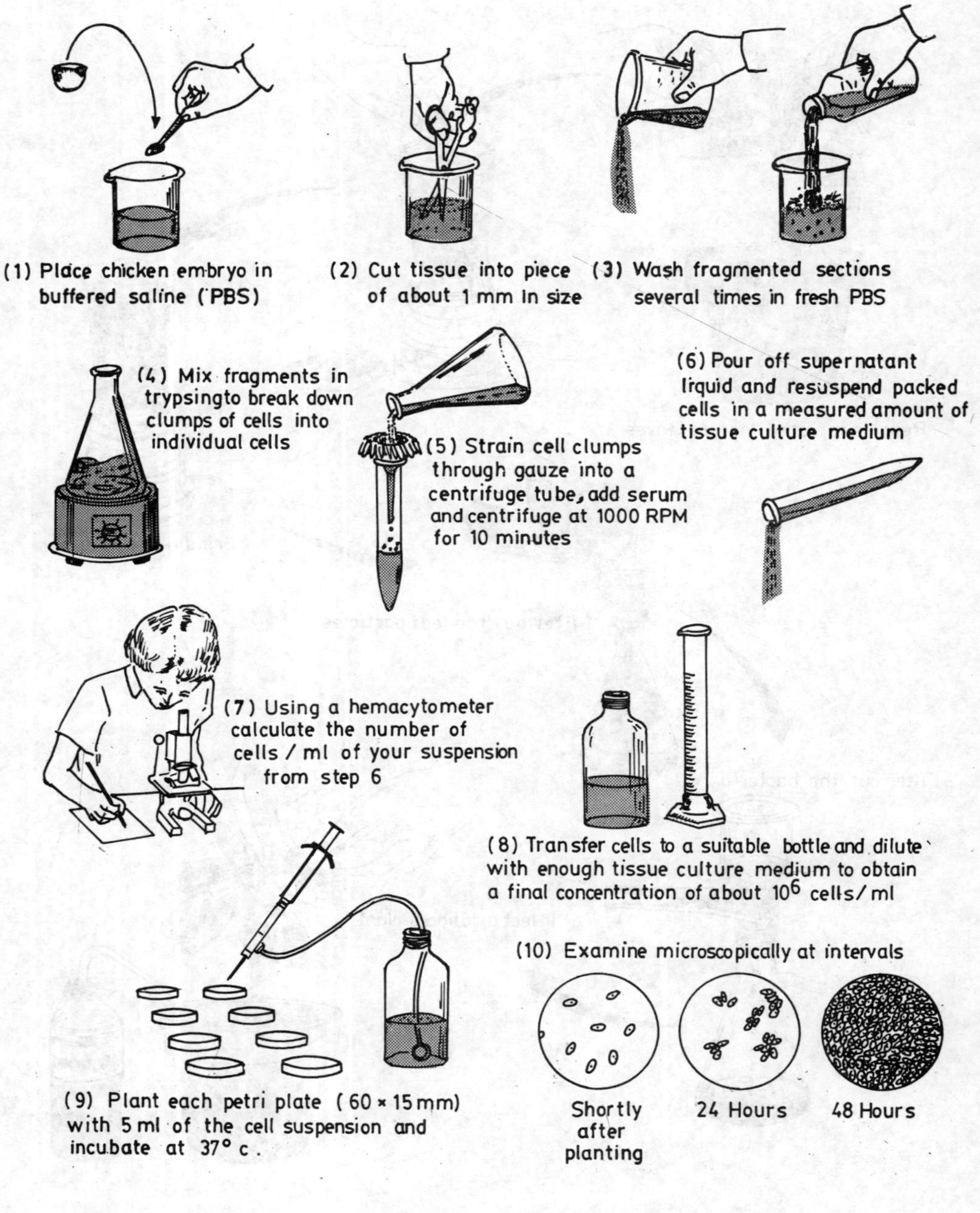

Fig. 24.6 Preparation of primary tissue for monolayer cultures.

4. Mix the fragments in Trypsin to break the clumps into individual cells.
5. Allow the cell clumps to pass through a gauze fitted on a centrifuge tube, add serum. and centrifuge at 1000 r.p.m. for 10 minutes.
6. Allow the clumps to pass through a gauze on a centrifuge tube, add serum and centrifuge at 1000 r.p.m. for 10 minutes.
7. Decant the supernatant and resuspend the cells in a measured amount of tissue culture medium.
8. Calculate the number of cells per ml of the above suspension using a Haemocytometer.
9. Transfer the cells to a flask and dilute it with enough tissue culture medium to obtain a final concentration of about 10^6 cells/ml.
10. Inoculate each petriplate with 5 ml of the cell suspension and incubate at 37°C.
11. Examine the plates microscopically at different intervals.
12. Aseptically aspirate the media from each of the tissue cultures, without damaging the monolayer.
13. Inoculate 0.5 ml of new castle disease virus to one monolayer and 0.5 ml of sterilised phosphate buffered saline to the other.
14. Rotate the cultures gently so as to mix the substances thoroughly.
15. Incubate the monolayers, keeping the cell side down at 37°C for 1 hour.
16. Add 5 ml of tissue culture medium having methyl cellulose to each monolayer, after one hour incubation.
17. Incubate the tissue culture for 2 days at 37°C.
18. Observe the monolayers under the microscope, watch for cytopathic effects in the inoculated one.
19. Aspirate the media from each monolayer.
20. Add 4 ml of 5% formalin to each monolayer culture (to preserve it) for 10 minutes Aspirate the formation.
21. Add 1 ml crystal violet (composition given in chapter 4) to each monolayer for 5 seconds. Aspirate the crystal violet. Allow the stained cells to dry and look for plaques in the monolayers.

6. Viral Haemagglutination

Red blood cells tend to agglutinate in the presence of antisera against them. Red blood cells can be easily enveloped with several antigens, the agglutination occurs in the presence of antisera against these antigens; the red cells act as passive carriers of the angtigens and aid the recognition of the antigen antibody reaction. The technique of using coated red-blood cells in this way is called Haemagglutination.

Here, the red cells are coated with a substance such as tannic acid which causes the antigen to adhere to them, than coated with the antigen. Dilutions of the antiserum are placed in

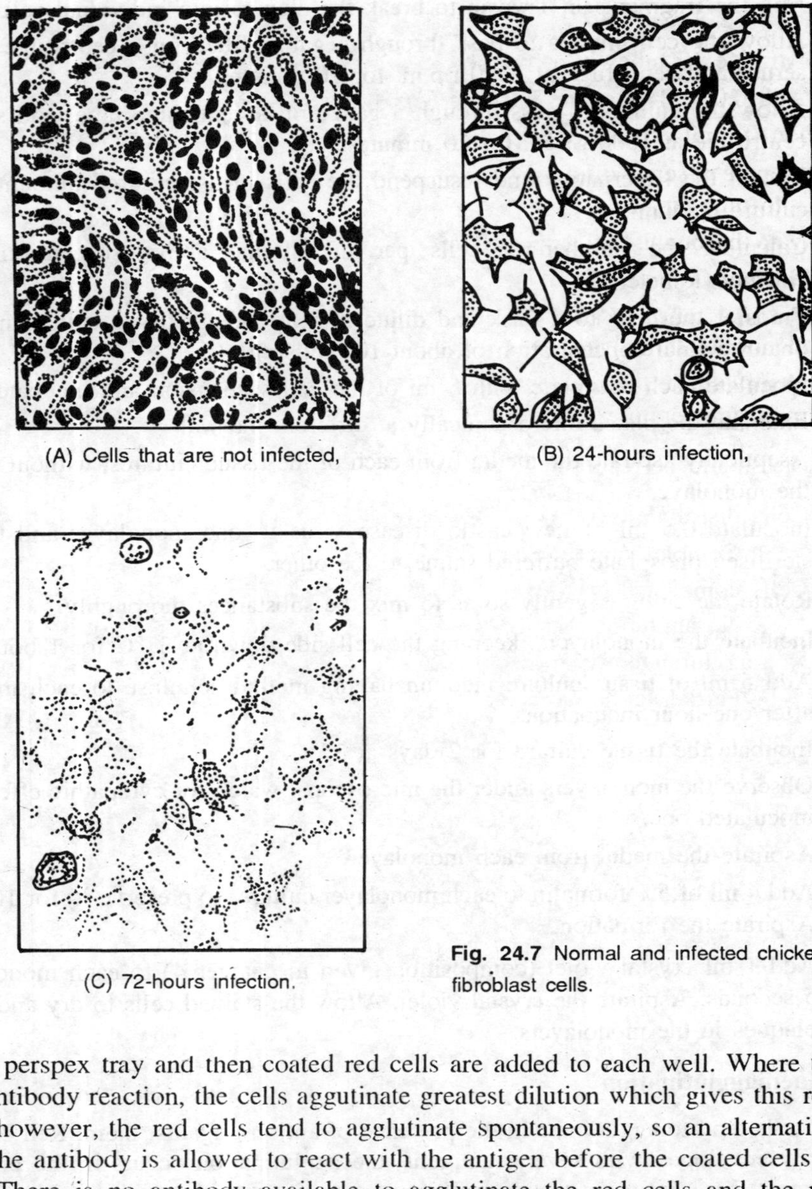

(A) Cells that are not infected,

(B) 24-hours infection,

(C) 72-hours infection.

Fig. 24.7 Normal and infected chicken fibroblast cells.

well in a perspex tray and then coated red cells are added to each well. Where there is an antigen antibody reaction, the cells aggutinate greatest dilution which gives this reaction. In practice, however, the red cells tend to agglutinate spontaneously, so an alternative method is when the antibody is allowed to react with the antigen before the coated cells are added is used. There is no antibody available to agglutinate the red cells and the reaction is inhibited. This technique is viral haemagglutination inhibition technique.

Another kind of haemagglutinations is brought about by the action of viruses which are responsible for mumps or influenza, cause red blood cells to agglutinate spontaneously and this agglutination is inhibited in the presence of specific antiviral antibodies.

Procedure

1. Arrange 12 tubes in a row and label as tube No. 1 to 12.
2. Add 0.05 ml of saline in tubes No. 2-12.
3. Put 0.1 ml of mumps antigen in the tube No. 1.
4. Prepare a serial dilution of antigen by taking 0.05 ml of antigen from tube No. 1 to tube No. 2.
5. Mix thoroughly and repeat the procedure till you have reached the tube No. 11.
6. Discard 0.05 ml of mixture from tube No. 11.
7. Add 0.1 ml of red blood cell suspension (1% suspension of chicken erythrocytes in normal satine) to tube No. 1-12.
8. Gently shake the rack to ensure each distribution of the cells, cover and leave at room temperature for 1 hr.

The tests should be recorded as follows :

1. Cells distributed evenly in the bottom of the well.
   Complete agglutination (+)
2. Cells compacted in a small red button at the bottom.
   No agglutination (–)
3. Non-agglutinated cells overlying a thin layer of agglutinated cells.
   Partial agglutination (+)